THE SCIENCE OF
ALIENS

THE SCIENCE OF
ALIENS

THE REAL SCIENCE BEHIND THE GODS AND MONSTERS FROM SPACE AND TIME

MARK BRAKE

AUTHOR OF *THE SCIENCE OF STAR TREK*

Skyhorse Publishing

*This book is dedicated to the most strange and beautiful alien
I know—the one who lives next door.*

CONTENTS

THE SCIENCE OF
ALIENS

INTRODUCTION

"So deep is the conviction that there must be life out there beyond the dark, one thinks that if they are more advanced than ourselves they may come across space at any moment, perhaps in our generation. Later, contemplating the infinity of time, one wonders if perchance their messages came long ago, hurtling into the swamp muck of the steaming coal forests, the bright projectile clambered over by hissing reptiles, and the delicate instruments running mindlessly down with no report . . . in the nature of life and in the principles of evolution we have had our answer. Of men elsewhere, and beyond, there will be none forever."
—Loren Eiseley, *The Immense Journey* (1957)

"Behind every man now alive stand thirty ghosts, for that is the ratio by which the dead outnumber the living. Since the dawn of time, roughly a hundred billion human beings have walked the planet Earth. . . . for every man who has ever lived, in this Universe there shines a star. But every one of those stars is a Sun, often far more brilliant and glorious than the small, nearby star we call the Sun. And many—perhaps most—of those alien Suns have planets circling them. So almost certainly there is enough land in the sky to give every member of the human species, back to the first ape-man, his own private, world-sized heaven—or hell. How many of those potential heavens and hells are now inhabited, and by what manner of creatures, we have no way of guessing; the very nearest is a million times farther away than Mars or Venus, those still remote goals of the next generation.

But the barriers of distance are crumbling; one day we shall meet our equals, or our masters, among the stars."

—Arthur C. Clarke, *2001: A Space Odyssey* (1968)

THE SCIENCE OF ALIENS

In 2005, I was a consultant to London's Science Museum. We were working on the Museum's *The Science of Aliens* exhibition, which was resident at the Museum in South Kensington for a brief period before touring venues around the world. Now, in the public imagination, the idea of aliens is mostly associated with the interstellar search that had absorbed astronomers for only the previous fifty years or so. Many potential Museum visitors would have known that, despite the best efforts of scientists, humans were yet to find convincing evidence of intelligent alien life in our Galaxy, or beyond.

Those of us working in the field were well aware of the fact that the science of aliens had a backstory far longer than the last fifty years, so at the Science Museum we were concerned with setting the story straight. Through a multimedia combination of artifacts, interactive exhibits, and audiovisual exhibits, we addressed the question "are we alone in the Universe?" by taking the long view—an approach that especially emphasized cultural and historical factors of the extraterrestrial life debate. Thus, *The Science of Aliens* exhibition borrowed from philosophy, film, and fiction, as well as the history of science, to portray the evolution of the alien debate over the last two and a half thousand years.

The thing is this: only by showing how scholars, moviemakers, and writers have devoted their energies to imagining life beyond the Earth can we do justice to a fascinating field which has often spoken poignantly about the human condition. Questions about life, the Universe, and everything, as Douglas Adams would have put it. And so our exhibition, and this book, pay as much attention to the science fiction archetypes as to what scientists can tell us about the real possibilities for alien life. Quotes from philosophy, film, and fiction sit alongside explorations of life on Earth and the extreme conditions in which it can survive. We look at missions to Moons and planets in our Solar System and what they can tell us about alien life, we go on to examine some exoplanets, and we conclude by looking at the prospect of communication with alien intelligence, the work of the Search for Extraterrestrial Intelligence (SETI) science program, and some of the messages sent out by humans into the Universe to try "talking" to extraterrestrial intelligence.

FROM YOUR LOCAL NEIGHBORHOOD TO LIFE IN THE UNIVERSE

For me, the journey to London's Science Museum began years before. In 1998, I had launched the first undergraduate module in Europe to examine the question of extraterrestrial life. The module, *Life in the Universe*, was a course in astrobiology, but one which straddled the boundaries between science, art, religion, and philosophy. The course provided an ideal starting point for discussions about the public and cultural placement of science. The media response to *Life in the Universe* was amazing. The ensuing media interviews on five continents were testament to this being the age of the alien. I went on to set up the world's first science and science fiction degree in 1999.

It's all a very strange affair, you might think, for an ordinary lad on the Celtic fringe of old Britain. What on Earth was so cosmic about Wales? What on Earth was so cosmic about any part of our little world in such a vast cosmos? When I started writing a history of the alien for Cambridge University Press, a local extraterrestrial link agreeably emerged. The first alien contact story published in the English language, it turned out, was written just twenty miles down the road from where I live. Back in the early seventeenth century, a local man named Francis Godwin wrote a tall tale called *The Man in the Moone*.

Now, this wasn't the first fantasy to come out of my country. We'd something of a famed medieval pedigree in the fantastic. But to my mind, *The Mabinogion*, revered as an almost national literary text, understandably focuses on the medieval lives of the nobility messing around on mountain tops. There was a sorry lack of space travel, and little mention of aliens. Godwin's book had both. It had space travel in the form of a trip to the Moon, and it had aliens in the form of, well, Moon people. Not only that, but his story had one of the best ever methods of propulsion through interplanetary space to the Moon: by geese! Godwin's protagonist, whose voyage of discovery goes astray to the Moon, rears and trains forty wild geese as a bizarre flying machine.

Francis Godwin was better known as the Bishop of Llandaff, and his book was the upshot of entertaining salty sea captains out of the ports of Bristol and Cardiff in an attempt to raise funds to restore the local cathedral. The Bishop started to write the book in 1589 but it was published posthumously in 1638 so he didn't live to see its meteoric success. *The Man in the Moone* went through two dozen editions, well into the eighteenth century, and was translated into many languages including French, Dutch, and German. Godwin's novel was seen as the archetypal space voyage for the next hundred years or so. When French dramatist Cyrano de Bergerac wrote his own space travelogue later in the century, *L'Autre monde ou les états et empires de la Lune*, he made sure his protagonist met Godwin's astronaut, one Domingo Gonsales. (Since Spaniards are able to navigate to new worlds on Earth, they must be equally capable of locating new worlds in space.) Even nineteenth-century writers such as Jules Verne and Edgar Allan Poe credited Godwin as a significant inspiration. But the high point of the influence of Godwin's Moon voyage story has to be with the freshly formed Royal Society. Godwin's story inspired them to consider the prospect of a space voyage to alien worlds.

ALIEN WORLDS

The search for life beyond the Earth can be fascinating in both terrestrial as well as extraterrestrial terms. My locality on this little world had had a hand in the very first plans to conquer outer space. There is, after all, something cosmic about most places on this planet. By 2005, I'd led the validation of the world's first degree in astrobiology, and the search for alien life. Meanwhile, mindful that local-born Terry Nation had invented those baddest of all aliens, the Daleks, BBC Wales regenerated *Doctor Who* for the twenty-first century. They transformed a scenery-shaking, self-parody of a television series into what the *Times* critic Caitlin Moran suggests is "despite being about a 900-year-old alien with two hearts and a space-time taxi made of wood, still one of our very best projections of how to be human." As I learned from all this work—creating university courses, consulting with the Science Museum, and writing books—stories about alien life, like *Doctor Who* and much of science fiction, have great potential for communicating subtle and soulful ideas about science. Gone are the days when alien fiction was just a geeky subculture—if it ever was.

Why not take a trip through the foggy ruins of time and read the compelling story of how the portrayal of extraterrestrial life has developed and evolved over the last two and a half thousand years? In Part I, we present a potted history of the major historical works of alien fiction and take a look at the evolution of the idea of alien life from the Bronze Age to the Space Age. In Part II, we focus on how explorations of life on Earth inform the debate about life beyond. In Part III, we look at alien worlds, how the planets and exoplanets we know about inform the way we think about fictional worlds. And finally, in Part IV, we ask more philosophical questions about our place in the Universe.

This is a story of history and adventure, science and invention, invasion and plunder, and some of the greatest movies of all time. It's a chronicle that sweeps continents and centuries, upending kings, cosmologies, religious dogma, and the dark age of faith. Following the alien debate traces momentous events that twice turned the world upside down: A medieval revolution that shifted the Throne of God to the far reaches of the Universe, and a Victorian revolution that struck at the heart of humanity itself. You never know, the story of *The Science of Aliens* might just start right outside your window.

PART I
ALIENS IN HISTORY

ALIENS IN ANCIENT GREECE

"Sailing the next night and day we reached Lamp-town toward evening, already being on our downward way. This city lies in the air midway between the Pleiades and the Hyades, though much lower than the Zodiac. On landing, we did not find any men at all, but a lot of lamps running about and loitering in the public square and at the harbor."

—Lucian, *A True Story*, translated by A. M. Harmon

THE WAR OF THE WORLDVIEWS

Alien fiction has some backstory. The tale of our changing view of alien life is one that begins during the grandeur that was the ancient Greek world. The brilliance of the Greeks will have a profound effect on the rest of our story. But what historical root does alien science have in antiquity? What roots in philosophy and fiction, and what flights of the imagination could possibly have happened over two thousand years ago? And what influence did these ideas have on the science and fiction of alien life that developed over the coming millennia? These are the questions this chapter answers.

Welcome to the war of the worldviews. There are two opposing philosophies in the Greek classical world that influenced succeeding ages, especially regarding astronomy and cosmology, and one of those philosophies is so prophetic about the existence of alien life that it seems truly modern. Even after ancient Greece, an evolving battle played out between conflicting philosophies, one we will trace in this book. At times of revolution, this battle peaked with great drama, like the clash between

Galileo and the Inquisition, and the controversy between Darwin and the creationists. But this schism is ancient in origin. Even in the ancient Greek world, thought diverged into two paths: one materialist, one idealist.

And so to the ancient world where we meet the first materialists in the Atomists and the first idealists in Plato and Aristotle. With the Atomists we find the shape of things to come: a worldview which embraced evolution, an atomic world of matter in motion with no God. With idealist thinkers Plato and Aristotle we find a philosophy of reaction, one that held no mortal life existed beyond the Earth.

THE ATOMISTS AND LIFE IN THE VOID

Ancient Greek science spawned the Atomists. They managed to weave a worldview which accounted for the creation of the cosmos and the way in which it worked. Given that this was two and a half thousand years ago, it's amazing that a material worldview, without recourse to gods and design, was divined at all. The Atomists believed in a Universe made of countless un-cuttable (*a-tomos*) particles. These particles moved through empty space, a movement that described all observable change. The atom was born. Naturally, their theory was far from perfect. The atoms were believed to be unalterable. But, after all, the Atomists lived in a time before gadgets, before the contraptions of discovery. Nonetheless, they understood that particles might explain nature's rich variety. And the Atomist worldview was one devoid of divine dabbling. They saw no need for gods.

Their other stunning innovation was the void: the empty space through which atoms moved. Injecting the notion of nothingness into science was a daring move. Earlier thinkers had thought the cosmos a plenum, a space entirely full of stuff. With what they regarded as common sense, most prominent philosophers loathed the idea of a vacuum. For example, in complete conflict with the Atomists, Aristotle was later to declare "nature abhors a vacuum."

DEMOCRITUS

In the West, one of the foremost Atomists was a Greek thinker of the fifth century BC named Democritus. Plato is said to have disliked Democritus so much that he wished his books burned. Democritus held that all things are in constant motion in the void and that there are innumerable worlds which differ in size. In some worlds there is no Sun and Moon, in others they are larger than in our world, and in others more numerous. There are some worlds devoid of living creatures or plants or any moisture.

Yet, for Atomists such as Democritus, this cosmic host of alien worlds was beyond reach. By the word "world," they didn't mean a different Solar System, like our own, that's potentially visible from Earth. Other worlds were not extrasolar systems, planets in orbit about distant stars. Rather, each was a self-contained cosmos, like our own, with an Earth at the center, and with planets and a stellar vault surrounding. Maybe these worlds could be called other "realms," a lot like our "multiverse" idea, a hypothetical set of other Universes, invisible and unreachable from our own Universe. These other worlds of the Greeks might be contemporaneous with their ancient world or may form a linear succession in time.

EPICURUS

Another famous Atomist was Epicurus. Born around 341 BC on Samos, Epicurus thought that there were infinite worlds both like and unlike this world of ours. He based this idea on what he believed to be the infinite number of atoms in deep space, so that there nowhere existed an obstacle to the infinite number of worlds. The logic is clear enough. As there must be an infinite number of atoms, and an infinite number of atoms could not have been exhausted by our finite world, then other worlds must be forged in the same way.

It's also clear that Epicurus believed aliens inhabited these infinite worlds. He thought that in all worlds there are living creatures and plants and other things that we see in our own world. Epicurus's worldview was also godless, naturally, as it follows on from a cosmos where atoms and void are the sole constituents, and all things that pass do so through chance collision of such atoms.

LUCRETIUS

A third famous Atomist was the Roman poet and philosopher, Lucretius. His famous first-century BC poem *On the Nature of the Universe* was an early exercise in science communication, a popularization of the ideas of

Epicurus. Lucretius's description of life in the cosmos includes the origin of species and is the longest and most detailed account from the ancient world. The main materialist and mechanist approach to nature was that of the Atomists such as Democritus and Epicurus. On the other side sat Aristotle and Plato and their idealist philosophy of design and purpose in the material world.

Lucretius was with the Atomists. His worldview accounts for the nature of the Universe from cosmology to the origin of life. As he says in *On the Nature of the Universe*, "Turn your mind first to the animals. You will find the rule apply to the brutes that prowl the mountains, to the children of men, the voiceless scaly fish, and all the forms of flying things. So you must admit that sky, Earth, Sun, Moon, sea, and the rest are not solitary, but rather numberless."

PLATO AND ARISTOTLE

The dominant cosmology of the ancient Greek world was that of Plato and his most famous student, Aristotle. These fourth- and third-century BC thinkers had the power and influence associated with being generally regarded as the two greatest figures of Western philosophy. For two millennia their cosmology held sway: a two-tier cosmos and a geocentric Universe. The Earth, mutable and corruptible, was placed at the center of a nested system of crystalline celestial spheres, from the sub-lunary to the sphere of the fixed stars. The sub-lunary sphere, essentially from the Earth to the Moon, was alone in being subject to the horrors of change, death, and decay. Beyond the Moon, the supra-lunary or celestial sphere, all was immutable and perfect. Crucially, the Earth was not just a physical center. It was also the center of motion, and everything in the cosmos moved with respect to this single center. Aristotle declared that if there was more than one world, more than just a single center, the elements of Earth and fire would have more than one natural place toward which to move—in his view, a rational and natural contradiction. Aristotle concluded that the Earth was unique. There was no room for the alien.

Plato went one step further. To the detriment of astronomy and the rational belief in life beyond the Earth, he fused math and theology and declared the planets divine. Their divinity was witnessed in their fixed and regular paths, orbits of perfect and circular motion. And, just as a finger gently rubbed along the rim makes a wine glass sing, so the divine planets sang in their circular paths about the central Earth. Change was banished from the heavens. The philosopher's highest calling, for Plato, became the consideration of perpetuity, his supreme pursuit the proof of man's immortality. For alien science, this meant oblivion. The developments in materialist philosophy now crumbled. The world of matter in motion, proposed by the Atomists, their Universe replete and seeded throughout with life, ground to a celestial halt.

We are now in a position to neatly summarize the contrasting views of the Atomists and Aristotelians, the followers of Aristotle:

Cosmos of Atomists	Cosmos of Aristotelians
Homogeneous	Hierarchical
Infinite	Finite
Random	Ruled by design
Purposeless	Teleological
Atheistic	Theistic
Cosmos is mutable and Earthly	Cosmos is immutable and perfect
Vacua exist	Vacua do not exist
Many worlds	Unitary Earth

Table 1. A comparison of the cosmologies of the Atomists and Aristotelians

A TRUE STORY: GREEK ALIEN FICTION

From those Aristotle-dominated days comes a work of speculative fiction that is just bristling with brilliant ideas about alien life. The tale is *A True Story*, and the author of the work is Lucian, an Assyrian satirist who wrote in Greek. Noted for his witty and scoffing nature, Lucian was one of the first novelists in the west, and *A True Story* the earliest known fiction about space travel and alien life. His tale contains themes of interplanetary imperialism and warfare, predating that kind of *Star Wars* space opera by many centuries.

Lucian's is a satirical work, written in prose, in which he parodies the kind of fantastic tales told by Homer in his *Odyssey* and other popular fantasies in Lucian's time. *A True Story* is a remarkable work. It anticipates many features of modern sci-fi, such as the fictional travelogues to Venus and the Moon, contact with alien life-forms, and wars between planets. And it was written seventeen centuries before the likes of H. G. Wells and Jules Verne practically "invented" the genre in more modern times. *A True Story* is, in short, the earliest work of alien science fiction.

The tale begins as Lucian and a company of adventuring heroes sail westward through the Straits of Gibraltar. They wish to explore the lands and peoples beyond the Ocean but are blown off course by a strong wind. In due course, they are lifted up by a giant waterspout, a rather ingenious form of liftoff, and set upon the Moon. His description of the Moon is fascinating. He says that from the narrator's point of view on the Moon, Earth looks just like another land "below," and is otherwise identical to any other alien planet.

Once on the Moon, they find a war raging, between the king of the Moon and the king of the Sun, over the colonization of the Morning Star. The armies which wage this war are staffed with the most exotic of militia. There are cloud centaurs, stalk-and-mushroom men, and acorn-dogs, "dog-faced men fighting on winged acorns." Tellingly, the Moon, Sun, stars, and planets are depicted as realistic backdrops, each with its own distinct habitat and peoples. Presently, the war is won by the Sun, and the Moon is clouded over. But not before more details of the Moon are unveiled: there are no Moon women, children being grown inside the Moon men.

This tale of Lucian's is remarkably replete with alien life and is an ancient example of "first encounter" fiction, stories in which humans come across alien life. One section of the story wittily describes men being seduced and then incorporated in a Borg-like manner by an alien life-form. Another section seems to suggest that the Moon people possess a well and a mirror, by which doings on Earth may be heard or watched respectively.

Lucian's *A True Story* is a revolutionary text concerning the Atomist ideas about life beyond the Earth. Astronomy is present in the trip to the Moon and beyond, and although his cosmology is unremarkable, it does tend to the Atomist, as it portrays the heavenly bodies as other worlds on par with our own. Lucian takes contemporary thought and uses it to picture alternative worlds in a way which makes his readers look at things afresh, forcing them to contemplate other worlds. Lucian's story is regarded by many as being the first true work of science fiction, the genre renowned for its depiction of alternative worlds, radically unlike our own, but similar in terms of scientific knowledge.

WORLDS BEYOND THE MOON

Lucian's alien worlds beyond the Moon are also remarkable. An almost endless procession of alien races seems to pour past. The doughty travelers reach the underworld of Tartarus, the first dystopia in science fiction. Here they briefly meet alternative human cultures in the form of pumpkin-pirates, nut-sailors, and dolphin-riders. Next come groups of alternate worlds that are consciously nonhuman in design. Landscapes composed entirely of wine or milk products suggest an alternate ecology. Nonhuman worlds inhabited by alien races are rational, rather than mythic, as they are scrutinized both in nature and anthropology. Witness the worlds of the vine-nymphs, the bullheads, a race of minotaurs, the ass-legs, a cannibalistic race of shape-changing women, and a race of feuding giants on floating islands. All aliens are portrayed as distinct from, but analogous to, Man.

Lucian's "Lamptown" creatures are even more remarkable. As the passage quoted at the top of this chapter shows, Lucian seems to have created an alternative and civilized alien life-form, which hints at the AIs of twentieth-century film and fiction. The product of human technology, not only do these lamp-creatures occupy "a city [which] lies in the air midway between the Pleiades and the Hyades," but Lucian even has a conversation with his own lamp: "I spoke to him and enquired how things were at home, and he told me all about them." His story ends with a promise that the next volume would feature adventures in this other world beyond the ocean, as if Lucian is suggesting he can go on generating these alien worlds without limit, and the world of the imagination—unlike the real world of late antiquity—remains forever open.

As an admirer of Epicurus, and like other Atomist philosophers, Lucian had a questioning mind about the wonderful variety in nature. In contrast to the idealists, Lucian was a thorough skeptic. He was very aware that the old myths were reemerging in the disguised form of some of the new philosophies and the degeneration of thought into pseudo-science. Lucian's tale is the most ancient example of what has come to be known as the intellectual nonconformism of science fiction. The wealth of estranged worlds in *A True Story* is the shape of things to come.

WORLD TURNED UPSIDE DOWN

"Whatever is born on the land or moves about on the land attains a monstrous size. Growth is very rapid. Everything has a short life, since it develops such an immensely massive body . . . In the course of one of their days they roam in crowds over their whole sphere, each according to his own nature; some use their legs, which far surpass those of our camels; some resort to wings; and some follow the receding water in boats; or if a delay of several more days is necessary, then they crawl into caves."
—Johannes Kepler, *Somnium*, translated by Edward Rosen (1634)

THE VOYAGES OF DISCOVERY

For nearly two millennia, Aristotle's cosmology held sway. But then came the ships. And with the ships, the great voyages across the world's oceans. Two Chinese inventions, the compass and the sternpost rudder, made long sea voyages viable, opening the seas to exploration, piracy, and war. An open ocean meant more accuracy: better observations, better charts, and better instruments, raising the need for a more predictive astronomy, a brand-new quantitative geography, and devices that could be used onboard ship as well as on land. But this need for better navigation had great consequences for science.

The great European sea voyages started with the Portuguese explorers around 1415 and opened the planet to capitalist enterprise. These voyages were the fruit of the first conscious use of astronomy and geography in the pay of glory and profit, for fledging empires soon realized that they were able to exert global control based on knowledge of territory: knowing *where* you were and *what* you owned. And so astronomy, navigation, and mapping became crucially important to trade.

THE SHIP AND THE TELESCOPE

The ship was key. As all the major trade routes were wet, commercial control very much depended on the speed and consistency with which long-range voyages took place along those wet routes. The discovery of the Americas and the quest for longitude and dominion in that New World was the impetus for a more aggressive approach to instrument making in science. And one of the first instruments of discovery was the ship itself.

The telescope too became a kind of ship. Consider this: pioneers of the spyglass took themselves, and their contemporary world as onlookers, to a place almost no one had ever imagined. But, like the destinations of a ship, the destinations of the telescope were also public knowledge. If you disbelieved what the likes of Italian astronomer Galileo claimed, you could take a look for yourself.

No one has put a better case for the ship as an instrument of discovery than twentieth-century German playwright Bertolt Brecht, in his greatest play, *Life of Galileo*. In the very first scene, Brecht has the Tuscan astronomer declare from his "humble study in Padua," "For two thousand years men have believed that the Sun and the stars of heaven revolve around them . . . The cities are narrow and so are men's minds. Superstition and plague. But now we say: because it is so, it will not remain so . . . I like to think that it all began with ships. Ever since men could remember they crept only along the coasts; then suddenly they left the coasts and sped straight out across the seas. On our old continent a rumor started: there are new continents! And since our ships have been sailing to them the word has gone 'round all the laughing continents that the vast, dreaded ocean is just a little pond. And a great desire has arisen to fathom the causes of all things . . . For where belief has prevailed for a thousand years, doubt now prevails. All the world says: yes, that's written in books but now let us see for ourselves."

So, the movement of the stars now had a cash value. But there was a problem in heaven. There had been little talk of the alien after Lucian.

From Plato and Aristotle on, few doubted the geocentric model. If their abstract theory didn't fit observations, common experience condoned it: you can see for yourself that the Moon, Sun, planets, and stars revolve around the central Earth! Philosophers and journeymen, poets and paupers, noblemen and kings all spoke of the Universe much as the ancient Greeks had described it. And laced into this conceit, since only Earth came accompanied with death and decay, was the idea that this was the sole place in the entire Universe where life resided.

THE COPERNICAN REVOLUTION

A new economy was born, and with it, the Renaissance. Open sea navigation led to booty, and plenty of it. A spirit of revolution was in the air. A conscious vanguard of merchants, scholars, and artists set about the task of constructing a new culture, capitalist in its economy, classical in its art, and scientific in its approach to nature. Initially, the plan had been to reconstruct the Classical World, returning to Plato and Aristotle. But soon the discovery of the New World made that ancient world seem unfashionable.

The main mission of the Renaissance was to rediscover and master nature. Astronomy became the acid test of progress. And its greatest achievement was a cosmology with the Sun at the center. The picture painted by Polish astronomer Nicholas Copernicus in *On the Revolutions of the Celestial Orbs* (1543) was of a rotating Earth in a Sun-centered system. Armed and inspired by newly edited Greek texts, a capable humanist could balance one Greek authority with another: Democritus against Plato, Epicurus with Aristotle. Equipped with a strong intellectual sense, and bags of bravery, such a humanist might even dare to center a new Universe around the Sun and worry later about shifting the Throne of God. And that is just what Copernicus did.

The alien arrived. Now, for the first time in written history, the Sun and planets are set out in their true heavenly order. AD 1543 is the date that marks one of the great turning points in the history of science. This infamous book, *On the Revolutions of the Celestial Orbs*, has the Sun at the center of its planetary system, a scandalous and heretical demotion of the position of the Earth to that of mere planet. Copernicus set in train a revolution in astrobiology. A new physics was born, and a new mantra: if the Earth is a planet, then the planets may be Earths; if the Earth is not central, then neither is humanity.

THE TELESCOPE

Next came the telescope. It was in March 1610 that Galileo's revolutionary discoveries with the spyglass were hurled like an incendiary device into a rather sleepy and dull academic world. Inside his brief but magnificent flyer, the momentous and now legendary book, *The Starry Messenger*, Galileo laid out the most incredible evidence for a new Copernican cosmos.

This book was a revolution in the communication of science. It was radical, not only in content but also in terms of its brief twenty-four pages. Remarkably, *The Starry Messenger* was Galileo's first scientific publication. It boasted of discoveries "which no mortal had seen before," and was penned in a new, tersely written way that few scholars had used before. The result was culture shock, arousing immediate and passionate controversy.

Consider the evidence, Galileo says in *The Starry Messenger*. In Aristotle's two-tier cosmos, only the Earth was meant to be subject to the horrors of change, death, and decay. Beyond the Earth, the celestial sphere was immutable and perfect. And yet through the spyglass, the picture is dramatically different. The Moon is pockmarked and bumpy. Far from crystalline and perfect. As for Aristotle's claims that the Earth is unique, Galileo shows that the Moon has Earth-like mountains, valleys, and hollows. What life lies there?

There was another decisive way in which these discoveries were revolutionary: you could see for yourself. In the age of faith, the enemy was monopoly. For centuries, knowledge had been shut up in the Latin bible, which only black-coated ministers were fit to interpret. Now, if you failed to believe what Galileo said of the lunar surface, anyone with enough ready cash to do so could pop out to the local spectacle maker, pick up a spyglass, and see the surface of the Moon for themselves.

A question in the nerves was lit. If the magnificent splendor of Aristotle's God-given cosmos is made especially for man's delight, why is it only through a machine, such as the spyglass, that man can justly savor its intricate parts and begin to know its true nature? What's more, the telescope had revealed a new and vast cosmos that stretched beyond the reach of human vision. It robbed the old Universe of its unity. The old cosmos was transformed.

Consider the Moon. With its prominent position in the night sky, it was the focus of exploration with the new "far-seer." And it was unique among the heavenly bodies in that its image through the spyglass unveiled features that made it easy to describe by terrestrial analogy. The Moon was evidence that there was material similarity between the Earth and the rest of the Universe. The Moon became the focus for the alien fiction of the age.

THE DARK SIDE OF MOON

Kepler's story *Somnium* is an extended voyage of discovery from the oceans of Earth to the dark seas of the Moon. Galileo's *The Starry Messenger* had inspired in Kepler early and uncannily prophetic ideas of alien life. In his first hasty letter to Galileo, *Conversation with the Starry Messenger*, Kepler hinted at his fascinating insight into Copernicanism, alien life, and space travel: "There will certainly be no lack of human pioneers when we have mastered the art of flight. Who would have thought that navigation across the vast ocean is less dangerous and quieter than in the narrow, threatening gulfs of the Adriatic, or the Baltic, or the British straits? Let us create vessels and sails adjusted to the heavenly ether, and there will be plenty of people unafraid of the empty wastes. In the meantime, we shall prepare, for the brave sky-travelers, maps of the celestial bodies—I shall do it for the Moon, you Galileo, for Jupiter." Amazingly, back in those days of Shakespeare, Galileo, and Kepler, the spaceship was first invented, inspired by the telescope.

Yet Kepler already had his own startling story to tell. The year before Galileo's Earth-shattering discoveries with the spyglass, Kepler had published in draft manuscript what many regard as the first ever work of modern science fiction. *Somnium* was a space voyage of discovery in the new physics that invented the alien and spoke of a cosmos that was soon to be unveiled by the telescope. Copernicus had shifted the hub of the Universe to the Sun. Kepler's aim was to explore the new and alien panorama from the alternative standpoint of the Moon. He wanted to show that a living being on the Moon would develop an astronomy that takes the Moon as the stable center of the cosmos, around which all else moves, orbiting Earth included. They would not only be likely to conceive of the cosmos as luno-centric and luno-static, but they would also witness a spectacle that to humans on Earth was so difficult to conceive: the Earth spinning around once every twenty-four hours.

In this way, the Copernican system could be compared and explored. The Moon becomes a mirror of faulty terrestrial science. The luno-centrism and luno-statism of lunar creatures matches the geocentrism and geostatism of narrow-minded Earth-dwellers. The Moon orbiting the Earth becomes a fictional device of the Earth orbiting the Sun. And novel observations of the heavens, and the Earth itself, are made viable.

Kepler also related how he read Lucian's *True Story* to teach himself Greek but was then gripped by this account of a voyage to the Moon: "He [Lucian], too, sails out past the Pillars of Hercules into the ocean and, carried aloft with his ship by whirlwinds, is transported to the Moon. These were my first traces of a trip to the Moon, which was my aspiration in later times."

SOMNIUM

Somnium is a truly radical work. Kepler's journey to the Moon is inventive genius. On arriving, Kepler's hero gazes down upon the Earth, noting its geography, its movement through space, and exploring the surface of the Moon and its alien inhabitants. Kepler's Moon is a truly alien world. His nightmare lunar vision begins with the deadly differences in temperature—blazing days, freezing nights—that plague the landscape. The starry canopy above the Moon is equally strange: across a bible-black sky, the stars, Sun, and planets scuttle relentlessly to and fro due to the Moon's orbit about the Earth. Indeed, this bewitching "lunatic" astronomy of Kepler's is yet to meet its match in fiction.

The aliens that stalk Kepler's Moon are not men and women, but creatures fit to survive the lunar haunt. Despite Galileo's holding an opposite view, Kepler imagined water on the Moon, hence the existence of life. Two centuries before Darwin, Kepler had already understood the link between life-form and environment: "Nothing on Earth is so fierce that God did not instill resistance to it in a particular species of animal: in lions, to hunger and the African heat; in camels, to thirst and the vast deserts of Syria; in bears, to the cold of the far north."

Kepler's approach is distinctly Darwinian. The *Somnium* narrative declares: "In plants, the rind; in animals, the skin, or whatever replaces it, takes up the major portion of the bodily mass; it is spongy and porous. If anything is exposed during the day, it becomes hard on top and scorched; when evening comes, its husk drops off." Kepler explains the inspirational source that fed his imagination in the creation of lunar life: "The precedents were our vegetables' and fruits' rinds, varying with nature's varying providence; the precedents were oysters' and turtles' shells, shaped like shields; the precedents were the calluses on the feet, the hoofs and the soles of animals."

The lunar nights are softened by the heat and light of the Earth, which stands still in the sky "as if nailed on." Some lunar territory is comparable to Earth's cantons, towns, and gardens. Like Earthlings, the Moonlings see the spectacle of the waxing and waning of the Earth's surface, fifteen times the apparent size of our Moon. And they gain respite from the constant cover of clouds, whose rain gives relief from the heat. The mountains of the Moon climb higher than those of Earth. Kepler's creatures attain a monstrous size. They feed only at night, for feeding after Sunrise often leads to death. The hide of the massive Moon serpents is permeable and, if exposed to the full force of the Sun, becomes seared and brittle. Such is life for this gigantic race of short-lived beasts. They bask for a fleeting instant in the rising or setting Sun, then creep into the impenetrable lunar darkness.

KEPLER, THE PIONEER

Somnium is a turning point. It signals the end of the old era and heralds loudly the dawning of the new science. Kepler's tale of alien life had an estranging effect on its reader, revealing the world in a new light. It is a fictional exercise in Copernicanism, as Kepler's tale does not only imagine life on our lunar world, reflecting the Moon's cosmic position and climate, it also assumes that lunar life will spawn out of the principle of environmental adaptation, as does life on Earth. The terrestrial and lunar worlds are analogous. Kepler's conclusion is Atomist: any part of the Universe that is potentially comparable to Earth in nature is designed to accommodate life.

The influence of Kepler's *Somnium* was huge. It motivated much later tales of interplanetary journeys, such as those of H. G. Wells and Arthur C. Clarke. And it inspired a potent motif for exploring the insignificance of man. Kepler was a pioneer of the vision of space as the home of many inhabited worlds. There is no better testimony to the power and imaginative sway of space fiction than this—despite the staggering odds against finding life beyond our planet, billions of dollars have been spent on sober scientific projects in search of alien intelligence. That search started with Kepler.

THE SYSTEM
OF THE WORLD

"There is the intolerable pride of human beings, who are convinced that nature was only made for them—as if it were likely that the Sun, a vast body four hundred and thirty-four times greater than the Earth, should only have been set ablaze in order to ripen their medlars and to make their cabbages grow heads!"
—Cyrano de Bergerac, *Les États et Empires de la Lune* (1657)

AND THE PLANETS MAY BE EARTHS . . .

Not everyone believed what they heard about Galileo and his discoveries with the "optick tube." His gift for vision promoted blindness in others. The crude nature of the mysterious gadget didn't help. But many were blinded by bias—they refused to look down the tube. A public controversy followed, similar in some senses to the UFO farce three hundred years later: allegations of optical illusions, haloes, and the unreliability of inexpert witnesses. Though Galileo and his telescope became the talk of the world's poets and philosophers, scholars in his own land were skeptical or blatantly hostile. Kepler's was the only weighty voice raised in defense.

Yet a revolution in science began with the discovery of these "alien" worlds. And the names of Kepler and Galileo have come to symbolize that revolution. They produced a map of the knowable just as the unknown was at the point of becoming known. Galileo's discoveries implied that Kepler's creatures may well be dwelling on the Moon. It was a critical new piece of evidence in the debate on the existence of alien life.

The Moon became real. In many ways, Galileo's telescopic discovery of the lunar surface rendered that alien world a real object for the first time. But at that very moment, we also feel a sense of wonder, or *estrangement*, from this new reality. By estrangement we necessarily mean a sense of flawed knowledge, the result of coming to understand what is just within our mental horizons.

This same sense of estrangement characterized the emerging alien fiction. Little wonder Galileo inspired Kepler to say the Jovian satellites (a word coined by Kepler himself) must be inhabited. It was Kepler who motivated H. G. Wells to write, nearly three hundred years later, "But who shall dwell in these Worlds if they be inhabited? Are We, or They, Lords of the World? And how are all things made for Man?"

The Moon was a touchstone. In thinking about the shadows on the lunar surface, Galileo had to assume they had similar causes to shadows on Earth, in order to understand the Moon's difference. And as eminent an astronomer as Kepler clearly needed to believe in aliens in order to make Galileo's discovery thinkable. Kepler understood that to know the Moon it was not enough to put telescopic observations into words. The words themselves had to be transformed by a new sort of fiction.

Throwing words at the Moon, as it were, has a dialectic effect—the words come back to us transformed. By imagining strange worlds, even those as commonplace as our Moon, we come to see science and life itself in a new perspective. Kepler's journey to the Moon was a very different piece of work compared with Lucian's *A True Story*. For one thing, it was written one and a half thousand years later. But the most striking difference was this: whereas Lucian's work is mostly fantasy, Kepler's *Somnium* was a conscious effort to address the new physics. And the crucial factor in the difference between the stories is the developed scientific consciousness evidenced by the newly invented telescope.

FICTION AND THE NEW PHYSICS

And so it began. The new breed of alien fiction was launched along with the new physics. Both marked the most massive paradigm shift that science has ever seen, as the old and cozy "Universe" of Aristotle slowly morphed into the new Universe of Copernicus: decentralized, infinite, and alien. The old Universe had the stamp of humanity about it. Constellations bore the names of Earthly myths and legends, and a magnificence that gave evidence of God's glory.

The new alien fiction was a response to the new and inhuman Universe. It dealt with the shock created by the discovery of our marginal position in a fundamentally inhospitable cosmos. Earth was no longer at its center, nor was it made of a unique material only to be found on terra firma. Nor was the Earth the only planet with bodies in orbit about it, as Galileo had discovered the Jovian Moons.

Kepler's *Somnium* was just the start. As alien fiction matured, its attempts to make human sense of Copernicus's new Universe became more sophisticated. And the fiction also played a growing part in the popularization of the new physics. The view through the telescope, however, was limited. The Moon, planets, and stars remained beyond reach. The spyglass, revolutionary though it was, did not bring the cosmos any closer. Indeed, it made the imagined journeys even more fantastic.

EMPIRES OF MOON AND SUN

Like Kepler's spaceships, the rocket made a surprisingly early appearance. Kepler had divined "Moon spirits" for his lunar journey. But in a single generation the space voyage quickly evolved into a rocket-propelled critical "contact" fiction, pioneered by Cyrano de Bergerac. Cyrano was, of course, notoriously unorthodox. French satirist and freethinker, his was a life immortalized by romantic legend. (Some readers may recall the 1987 movie *Roxanne*, starring Steve Martin. The film was a modern retelling of the 1897 verse play *Cyrano de Bergerac*.) But the real-life Cyrano equaled the legend of the swashbuckling swordsman with a large snout. Cyrano allegedly fought with the elite Gascon Guard and was a radical who liked to parade his heterodoxy in extravagant style.

Cyrano had studied with French philosopher Pierre Gassendi, whose life's work was to change the public perception of science. Gassendi was a canon of the Catholic Church who tried to reconcile Epicurean atomism with Christianity. He had a reputation as a successful communicator of science, a reputation that owed much to the influence of his famous pupil. And Cyrano's repute rested on a trilogy of tales, known collectively as

L'Autre Monde (*Other Worlds*) and separately as *Les États et Empires de la Lune* (*The States and Empires of the Moon*), *Les États et Empires du Soleil* (*The States and Empires of the Sun*), and the lost work, *L'Histoire des Etoiles* (*The History of the Stars*).

In his *Lune* and *Soleil*, Cyrano makes an Atomist argument for an infinity of inhabited alien worlds. As well as the influence of his mentor Gassendi, Cyrano was also swayed by Descartes. Descartes was famed for the Atomist cosmological theory of vortices, which said that the cosmos was filled with matter in various states, swirling in vortices about stars like the Sun. Unlike Descartes, Cyrano radically assumed the existence of as many Copernican planetary systems as there are Suns. Though Cyrano was writing in fiction, rather than fact, his theories nonetheless comprise a system of the world. The system is presented mostly in the form of conversations in Cyrano's narratives, where his hero encounters inhabitants of both Moon and Sun, in the most imaginative and mischievous tales.

Cyrano's imagination captured the mood of the age. Like Lucian, Cyrano's imagined alien life was openly atheist. Copernicus, Galileo, and Kepler had been pious men. Not Cyrano. Under the influence of Descartes, Cyrano banished God from the heavens, with pure reason as his guide. His alien tales were remarkably popular all over Europe, and an influence on writers such as Jonathan Swift, Edgar Allan Poe, and Voltaire.

Cyrano took great delight in scandalizing the Church. The first two volumes of his trilogy were so strident that they had to be toned down in their sacrilege and heresy. The secular culture implied by the new physics was quite openly expressed in Cyrano's alien fiction. The notion of a cosmos fit for life was purged of any divine nuance. Cyrano's alien worlds were not the work of some Creator. And his alien fiction was a unique attack on human self-esteem and hubris. Kepler had confronted the Copernican cosmos in fiction. Now, and with comparable daring, Cyrano tackled the topic of Cartesian physics, a non-Christian view of an alien Universe, which Descartes himself shied away from in public.

L'AUTRE MONDE

Books one and two of *L'Autre Monde* told tales of imagined meetings between Man and alien. Kepler had presented a Moon world in *Somnium*, which was at least in some ways based upon observation and number. Cyrano does no such thing. Cyrano's fictions of the lunar and solar worlds make little pretense that they are based on observational reality. Rather, they are fictitious, off-world explorations of arrogant and narrow-minded objections to the existence of alien life.

The Universe of *L'Autre Monde* is Atomist. A cosmos is conjured up, one in which there are as many planets as stars, and these numberless worlds, like the Moon and Sun in his stories, are all peopled by rational creatures. An intelligent Universe is only the beginning. Cyrano uses this cosmic landscape to cast doubt over the nature of Man. His conclusions are not reassuring. The books are witness to the most critical view of the alleged superiority of Man, and of religion, than any other work of its time.

Cyrano's voice is crystal clear in *The States and Empires of the Moon*. For one thing, the narrator is named Dyrcona, an anagram of d[e] Cyrano. And for another, what unfolds is nothing short of a lunar lampoon of hubris on Earth. The narrative begins, ingeniously enough, with space travel by bottled dew!

Take one: by fastening the bottles about his waist, the evaporating dew lifts Dyrcona to the Moon. But his mission fails, and he falls to Earth, initially to the French colony of Canada. Once captured, the Viceroy, who is very interested in Dyrcona's means of travel and the new astronomy, questions Dyrcona. He replies, "I believe that the planets are worlds surrounding the Sun and the fixed stars are also Suns with planets surrounding them." The Viceroy becomes a Copernican and concludes that the cosmos is vast, and Aristotle wrong.

Take two: by fixing fireworks to a traveling machine, Dyrcona now rockets to the Moon. In the opinion of twentieth-century science fiction author Arthur C. Clarke, Cyrano invented the ramjet, a form of air-breathing jet engine, which uses an engine's forward motion to compress incoming air. Though the invention of the ramjet is traditionally attributed to French inventor René Lorin in 1913, Clarke drew attention to this innovative passage by Cyrano, written around two hundred and fifty years before

Lorin: "I foresaw very well, that the vacuity . . . would, to fill up the space, attract a great abundance of air, whereby my box would be carried up; and that proportionable as I mounted, the rushing wind that should force it through the hole, could not rise to the roof, but that furiously penetrating the machine, it must needs force it upon high."

The Moon that Cyrano unveils is another world, a lunar landscape stalked by a race of giants. It is a case of four legs good, two legs bad, as the lunar men, the Lunarians, walk on all fours, and greet Dyrcona with some disdain—a little biped from their "moon." The lunar beasts too walk on all fours, making Dyrcona's appearance look all the more alien. On Cyrano's Moon, humans are not as fit as beasts, let alone the Lunarians. Dyrcona encounters a Phoenix and an alien intelligence in the form of talking trees. He learns that rather than burn in his nest to birth a new bird, the Phoenix flies off to the Sun to join the community of birds there. The trees maintain that the reason and soul within them are their own, and not the result of humans morphed into trees. They are truly alien.

The medieval Church suffers a lunar lampoon. Scriptural claims that only man has the faculties of reason and immortality are made to look ridiculous. After a dazzling description of an atheist genesis of life in the Universe, and a committed declaration that the stars are inhabited, Cyrano explains that reason is the keystone of the enlightened lunar culture. In this culture, death is the fulfillment of life, and man's belief in God is considered a deficiency of reason. The decidedly intelligent Lunarians feed on scents and sleep on flower blossoms.

Humanity's place on the cosmic ladder is also the rub in a run of weird events. The Lunarians train Dyrcona to perform tricks, just like chimps on Earth. He learns that the lunar men think of humans as baboons. They're kept as ludicrous pets and shown in a lunar zoo. Indeed, the discussions that Dyrcona has while at the zoo are especially amusing to the Lunarians. But when the rumor starts that the caged biped may be a flawed human, the lunar church intervenes. A law is decreed: any suggestion of similarity between the Earthly creatures and Lunarians, or even lunar beasts, is blasphemy.

Finally, Dyrcona stands trial. The charge against him: believing his world is central, and not a "moon" of the Moon! The lunar men argue that humanity is completely insignificant. The bipeds are considered monsters devoid of reason, best thought of as "plucked parrots." Paradoxically, the court still decides to declare Dyrcona a "man" so that he can be charged. His humiliation is to publicly recant his geocentric beliefs on every street corner of the Moon.

DARWIN AMONG THE ALIENS

"No one would have believed in the last years of the nineteenth century that this world was being watched keenly and closely by intelligences greater than Man's and yet as mortal as his own ... With infinite complacency men went to and fro over this globe about their little affairs, serene in their assurance of their empire over matter ... At most, terrestrial men fancied there might be other men upon Mars, perhaps inferior to themselves and ready to welcome a missionary enterprise. Yet, across the gulf of space, minds that are to our minds as ours are to the beasts that perish, intellects vast and cool and unsympathetic, regarded this Earth with envious eyes, and slowly and surely drew their plans against us."

—H. G. Wells, *The War of the Worlds* (1898)

THE ORIGIN OF SPECIES

By the nineteenth century, material progress was realized. Science had secured its dominion over nature. Newton's system of the world was set free. The *philosophical* engine, the early steam engine, drove locomotives along their metal tracks; the first steamships crossed the Atlantic; the great transport magnates were building bridges and roads; telegraphs ticked intelligence from station to station; the Lancashire cotton works glowed by gas; and a clamorous arc of iron foundries and coal mines powered this Industrial Revolution.

Isaac Newton had created a clockwork cosmos, a mechanical worldview. As the machinery began to mesh, science encroached upon all aspects of life, meeting every challenge with a different invention. Progress and technology seemed inseparable. The machines of science were devised not merely to explore nature, but to exploit it. For every factual gadget, fiction spawned a thousand visions.

Yet the great discovery of this machine age, in which Darwin played a part, was in biology. This discovery described life's evolution. Evolution had been argued before, so the true innovation of the age was to discover the evolutionary mechanism by which new species came to be. For no small reason did Darwin name his 1859 book *The Origin of Species*. "Evolution" was not a word Darwin liked to use. "One may say," Darwin wrote, "there is a force like a hundred thousand wedges trying to force every kind of adapted structure into the gaps in the economy of nature, or rather forming gaps by thrusting out weaker ones."

Here was a natural mechanism that was not only global, but also potentially cosmic. It was a means that canvassed continually to snuff out most variations, while keeping those few carried by individuals who had won the struggle to survive and breed. Here was "natural selection." The individual differences between members of a species, coupled with environmental factors, shape the chances that an individual will pass its characteristics on to posterity.

Darwin's *The Origin of Species* identified natural selection as the mechanism that creates new species. Twenty-three centuries after the Atomists suggested that all was in a state of flux, Victorian science resurrected an Atomist world in a state of constant change. The new science had identified two things. One, that nature favored variety. And two, that nature prefers geographical spread, for the further afield a species is scattered, the less tied their prospects to a single setting.

Of course, the theory of evolution is not the result of a single worker. From antiquity on, the finest philosophers pondered the rich variety of life. And over those centuries, divine creation had not always been thought of as the causal factor. A thread of eminent thinkers, from Epicurus and Lucretius to Leonardo da Vinci, had preferred more secular ideas. Rather than believing that every form of life was created by God and has stayed unaltered since, these sages looked to nature's inherent patterning for an answer.

INCENDIARY EVOLUTION

Darwinism struck at the heart of humanity itself. Newton's system of the world essentially reestablished the integrity of design, which had been shattered by Copernicus and Galileo. The Christian picture of creation had stayed more or less untouched. Man was still made in the image of God. After Darwin, the book of Genesis lay in shreds as a literal history.

The Origin of Species was appropriated by the radical, anticlerical wing in politics, and twisted to its agenda of *laissez-faire* capitalism. Darwin unwittingly provided an alibi for brutal exploitation by the "fittest," the subjugation of lesser by higher peoples. Association with nature "red in tooth and claw," as Tennyson put it, could justify even war itself. The notion of the "chosen ones," that one-time apology for the supremacy of classes or races, had withered. It was replaced by a "Darwinian" validation of a brave new world of reason, industry, and empire.

Darwinism injected the lifeblood of history into science. "He who . . . does not admit how vast have been the past periods of time may at once close this volume," Darwin wrote in *The Origin of Species*. The theory of evolution could have been used to unite the human and nonhuman spheres. Instead of such an emphasis on the affinity of Man and nature, the social evolution of humanity was eclipsed by scientism. A limited science focus resulted in the perverse justification of race theories and imperialism. And the utopian tales of the future that followed in Darwin's wake explored the connection between nature and society.

H. G. WELLS ON ALIEN SPACE

H. G. Wells was not the first Victorian author to explore Darwinian notions of alien space. But Wells's 1898 alien classic, *The War of the Worlds*, is arguably the single greatest work of alien fiction. Herbert George Wells was born into an English lower middle class that had previously produced another key author in Charles Dickens. Wells's mother had been in service, his father a gardener. Though optimistic of becoming socially mobile as shopkeepers, the shop failed year after year. Wells's employment began as apprentice to a draper. But it ended rather sharply when he was told he was not refined enough to be a draper. Such rejection at the sharp end of a class-conscious society became one of the key influences on Wells's critique of the Victorian world.

Wells's defining moment in science came on meeting "Darwin's Bulldog," the great T. H. Huxley. Wells was bright enough to have won a scholarship to the Normal School of Science, later the Royal College of Science, where he studied evolutionary biology under Huxley. A fervent Darwinian, Huxley was the chief science communicator in England at the time. He had introduced the word "agnostic" into the lexicon, and impressed Man's simian ancestry on the public imagination. His public lectures attracted massive audiences. Two thousand were reportedly turned away at St. Martin's Hall in 1866, the year of Wells's birth.

THE WAR OF THE WORLDS

The mood of the age was one of fear. Between 1870 and the start of WWI there were hundreds of authors writing invasion literature, often topping bestseller lists in many countries of Europe and the United States. This pervading sense of fin de siècle was fed limitless power by Darwinian evolution. It allowed Wells to produce work whose appeal has dimmed little since.

Into this fragile climate marched an alien invasion. Starting with the most superb opening in the entire history of science fiction, quoted at the top of this chapter, Wells's *The War of the Worlds* does for alien fiction what his 1895 book *The Time Machine* had done for time. Wells's Martians are agents of the void. They powerfully shatter the human domain. Wells's book is the finest and most influential of all alien contact fiction. It is the first Darwinian fable on a universal scale.

Wells's wrath was directed at the idea of the "becoming of Man." The book begins with a quote from Kepler: "But who shall dwell in these Worlds if they be inhabited? Are We, or They, Lords of the World? And how are all things made for Man?" It is a battle between Earth and Mars, humans and Martians. The narrator of this struggle for survival is a philosopher, writing a thesis on the progression of moral ideas within civilization. His assumption of a bright future is rudely blown apart mid-sentence by the brutal natural force of evolution in the shape of the Martian attack.

The War of the Worlds features the first "menace from space." Regardless of Kepler's lunar serpents, the modern alien owes everything to Wells. With their distinctive physiology and intellect, the Martians are the archetypal alien. Wells's story was first serialized in 1897. Other space novels at the time, especially those about Mars, spoke of travelers journeying through the Solar System on some kind of imperial and Victorian conquest. The climax of their meeting with aliens was invariably a sense of triumphant superiority, garnered with an enhanced human self-esteem. In Wells's novel, however, it is the Martians who conquer space. They land on Earth and pose a serious threat. The fear is real and gripping.

Wells had been keenly aware of the possibility of alien life for some time. He had contributed to the extraterrestrial debate in 1888 at the Royal College of Science on the subject "Are the Planets Habitable?" He had also written essays in support of fellow pluralists Kepler, Camille Flammarion, and Percival Lowell, whose mistaken obsession with "evidence" of life on Mars had recently reached Europe. Rather than being a whimsical work of fiction, *The War of the Worlds* demolishes the idea that Man is the peak of evolution. Instead, Wells creates the myth of a technologically superior alien intelligence.

MARTIAN IMPERIALISM

Mars is a dying world. Its seas are evaporating, its atmosphere melting away. The entire planet is doomed so "to carry warfare Sunward is, indeed, their only escape from the destruction that generation after generation creeps upon them." Thus, the terror of the void is brought down to Earth. Wells delivers repeated reminders of "the immensity of vacancy in which the dust of the material Universe swims" and evokes the "unfathomable darkness" of space. Life is portrayed as precious and frail in a cosmos that is essentially deserted.

The book consciously conveys the quality of the void—immense, cold, and indifferent—in its rendering of the aliens. But it is the Martian machines that vividly hammer home the cosmic chain of command: "It is remarkable that the long leverages of their machines are in most cases actuated by a sort of sham musculature . . . Such quasi-muscles abounded in the crablike handling-machine . . . It seemed infinitely more alive than the actual Martians lying beyond it in the Sunset light, panting, stirring ineffectual tentacles, and moving feebly after their vast journey across space."

The Martian Tripods tower over men in the flesh, as the vast intellects of their occupants tower over human intelligence. Physically frail, but mentally intense, the Martians and their superior machines are instruments of human oppression. Their weapons of heat rays and poison gas are dehumanizing devices of mass murder. All efforts at contact are futile, furthering the idea of the aliens as an unrelenting force of the void.

Understandably, the human reaction to this cosmic struggle is alienation. The narrator alienates himself from the grim reality of the inevitable triumph of death over life: "I suffer from the strangest sense of detachment from myself and the world about me; I seem to watch it all from the outside, from somewhere inconceivably remote, out of time, out of space, out of the stress and tragedy of it all."

WELLS'S ARCHETYPAL ALIEN

Wells planned the invasion on bicycle. It is pleasing to picture him, as he says in his autobiography, mapping mayhem as he "wheeled about the district marking down suitable places and people for destruction by my Martians." As early as 1896 he declared his intentions to "completely destroy Woking—killing my neighbors in painful and eccentric ways—then proceed via Kingston and Richmond to London, which I sack, selecting South Kensington for feats of particular atrocity." It is this exquisite violence of Wells's imagination that marks his genius.

As Wells had planned on bicycle, so it was in fictional South Kensington. The area is haunted by the sound of the Martians howling "Ulla, ulla, ulla, ulla." The alien invaders have finally fallen prey to an Earthly predator—microbes. The fate of the Martians emphasizes not only the insignificance of human resistance to the struggle, but also the latent power of unsolicited natural selection. Evolution kills.

Wells's seminal alien story ends with the dialectic of the Martians. On the one hand, the narrator feels for the aliens. The tragedy of the Martians is the tragedy of Man. The aliens' torment speaks strongly of the common struggle for all life in a hostile Universe. On the other hand, Wells has constantly portrayed the Martians as vast, cool, and unsympathetic. Alien in tooth and claw, as it were. The narrator may merely be projecting emotion onto creatures that are fundamentally inhuman.

Wells's Martians are a fascination. They are alien, yet they are human. They are what we may one day become, with their "hypertrophied brains and atrophied bodies," the tyranny of intellect alone. But they are also political. Wells evidently has the Martians brutally colonize Earth, but "before we judge them too harshly we must remember what ruthless and utter destruction our own species has wrought . . . upon its own inferior races . . . Are we such apostles of mercy as to complain if the Martians warred in the same spirit?"

Finally, in writing of the alien, Wells is also writing about his own Victorian machine world. The Martians are a veiled criticism of the machine age, with its application of science to industry. In this fashion, *The War of the Worlds* condemns the social machine of capitalism, the reducing of humans to anonymous cattle, the indifference at any attempts to communicate the inhumanity of the system and the feeling of alienation.

It is a Wellsian notion of time as the fourth dimension from *The Time Machine* that we shall find in the next chapter. It is the fabric of a Universe of effortlessly wheeling Galaxies in gently curving space-time. Into this bible-black sky, alien fiction now plunged, trading the Wellsian terror of the void for a freedom of infinite space-time.

EINSTEIN'S SKY

"We regard the cosmos as very beautiful. Yet it is also very terrible. For ourselves, it is easy to look forward with equanimity to our end, and even to the end of our admired community; for what we prize most is the excellent beauty of the cosmos . . . In our Solar System there are the Martians, insanely and miserably obsessed; the native Venerians, imprisoned in their ocean and murdered for Man's sake; and all the hosts of the forerunning human species. A few individuals no doubt in every period, and many in certain favored races, have lived on the whole happily. And a few have even known something of the supreme beatitude."
—Olaf Stapledon, *Last and First Men* (1930)

"Cosmic pluralism, the plurality of worlds, or simply pluralism, is the belief in numerous 'worlds' in addition to Earth (possibly an infinite number), which may harbor extraterrestrial life."
—Mark Brake, *Alien Life Imagined* (2012)

MAN'S PLACE IN THE COSMOS

Earth was an alien planet. By the dawn of Einstein's century, Earth had been alien for some time. The paradigm shift of the Copernican revolution cut both ways. Not only did the revolution make Earths of the planets, but it also brought the alien down to Earth. The old "Universe" of Aristotle had been small, static, and Earth-centered. It had the stamp of humanity about it.

The new Universe was inhuman. The deeper the telescopes probed, the darker and more alien it became. "The history of astronomy," suggests novelist Martin Amis, "is a history of increasing humiliation. First the geocentric Universe, then the heliocentric Universe. Then the eccentric Universe—the one we're living in. Every century we get smaller. Kant figured it all out, sitting in his armchair . . . The principle of terrestrial mediocrity."

American astronomer and science fiction author Carl Sagan went further, suggesting that humans suffered a series of "Great Demotions" in the last five centuries. First, consider Earth. It was not at the center of the Universe. Nor was it the only object of its kind, made of a unique material only to be found on terra firma. Next, consider the Sun. Not at the center of the Universe, not the only star with planets, nor eternal. If cosmology could be endured, there was biology. Man now lived among the microbes. He had no special immunity from natural law, and there was vanishingly little evidence of a divine image. Each successive demotion had a huge impact, both on the human condition and on the meaning of life in the cosmos.

Then, there was Einstein's sky. As the twentieth century unfolded, astronomers would uncover a cosmos composed of more stars than grains of sand on all the beaches of planet Earth. The infant century was about to boast a vast Milky Way Galaxy, not at the center of the cosmos, nor the only Galaxy within it. Two trillion other Galaxies have since been discovered, adrift in an expanding Universe so immense that light from its outer limits takes longer than twice the age of the Earth to reach terrestrial telescopes. And there may be other Universes.

This chapter considers the impact of Einstein's sky on the alien fiction of the early twentieth century. We shall be mindful, as the story unfolds, of the words of "Darwin's Bulldog," English biologist T. H. Huxley: "The question of questions for mankind—the problem which underlies all others, and is more deeply interesting than any other, is ascertainment of the place which Man occupies in nature and of his relations to the Universe of things." Keep in mind Carl Sagan's view that the final demotion would be the discovery of another biological intelligence in the Universe. We shall see, as the centuries draw on, that the stage upon which this debate plays out grows larger and larger.

ANTHROPOCENTRISM

Let's go back to the geocentric Universe that was to emerge as the dominant cosmology of the ancient Greek world. Even meticulous observations of the night sky by our ancestors led to a misplaced sense of the importance of Man. Plato's idealist perspective was one of a divine cosmos, a Universe in which change was cast out. The crisis that faced the ancient Greek world bled into its physics, as the philosophers of the day tried to save society from the sociopolitical disaster into which it was descending.

As it was on Earth, so it shall be in the heavens. In fear of change, most Greek scholars believed that the divine nature of the cosmos shone through in the unchanging and regular paths of the planets with orbits of perfect and circular movement. In Aristotle's vision, the "Universe" was geocentric but, for him, our world was not just a physical center—it was also the center of motion. Everything in the Universe moved with respect to this single center. Aristotle said that if there were more than one world, more than just a single center, elements such as Earth would have more than one natural place to fall to. Aristotle essentially declared the Earth unique, and Man alone in the Universe.

The medieval sky too was Aristotelian and geocentric, but this vision was utterly transformed by the Church. Through Christian symbolism, Aristotle's Universe of spheres mirrored man's hope and fate. Both bodily and spiritually, man sat midway. His crucial position was halfway between the inert clay of the Earth's core and the divine spirit of the heavens. And though the only known and observed life in the cosmos was here on Earth in filth and uncertainty, close to Hell, Man was always under the eye of an assumed God. The choice was whether to follow a human and Earthly nature down to its natural place in Hell or to engage with the spirit and follow the soul up through the celestial spheres to God. Any alteration in the great plan of Aristotle's "Universe" was bound to corrupt this drama of Christian life and death. To shift the Earth was to sever the continuous chain of created being, and to move the Throne of God himself.

Nonetheless, shift it we did. Heliocentrism was born in the Copernican revolution, and with it a new tradition in the plurality of worlds. The Copernican worldview struck at anthropocentrism. If the Earth is a planet, then the planets may be Earths; if the Earth is not central, then neither are we. Copernicus himself was silent on alien life. But there was a new sense of wonder at the emerging possibilities. And this sense of wonder was responsible for the birth of science fiction.

ALIEN FICTION

To recap, science fiction began with the revolution of Copernicus. It was born along with the paradigm shift of the old Universe into the new. Aristotle's cozy geocentric cosmos was anthropocentric. The new Universe of Kepler and Galileo was, at least potentially, decentralized, infinite, and alien. In many ways, the early alien fiction can be read as a response to the cultural shock created by the discovery of humanity's marginal position in a Universe fundamentally inhospitable to Man.

Science fiction embraced the alien. It emerged as our attempt to make human sense of Copernicus's new Universe. We saw this in Kepler's invention of the alien in *Somnium*. The aliens that stalk *Somnium*'s Moon world are creatures fit to survive a strange habitat. It was the first alien fiction of the age. But the alien voyages evolved quickly, and became a potent way for exploring the insignificance of Man.

Cyrano de Bergerac's *L'Autre Monde* trilogy was among the first to use satire in its portrayal of man's place in the cosmos. In a series of weird events on the Moon and Sun, Cyrano addresses anthropocentrism head-on by exposing the inferiority of humans in an alien setting. On the Moon, his antihero stands trial for believing that his own planet is a world, not a "moon." The Lunarians believe in the total insignificance of humans, as their human counterparts might do on Earth. The bipeds are thought to be monsters devoid of reason and are best described as "plucked parrots."

Cyrano's travels to the Sun told a similar tale. Man is portrayed as arrogant, a deluded creature who believes the entire animal world, and the environment, to be at his disposal. But the solarian birds are equally arrogant. They too believe themselves the most supremely rational and cultured beings in the cosmos. In this way, Cyrano satirized the bigotry and beliefs of the age. *L'Autre Monde* was a secular picture of the potential meaning of a Universe fit for life. It was a Copernican stand against the anthropocentric notion that Man was the center of creation.

GALACTOCENTRISM

Geocentrism gave way to heliocentrism, which yielded eventually to galacto-centrism. German-born British astronomer William Herschel was sweeping the night sky with state-of-the-art handmade telescopes of the eighteenth century. Mid-century, Immanuel Kant proposed the idea that the fuzzy patches of light, known as nebulae, were actually distant "island Universes" composed of myriad star systems. The form of our Milky Way Galaxy was expected to be similar to the nebulae.

By 1783, two years after he discovered Uranus and doubled the size of the Solar System, Herschel began his attempts to gauge the size and scale of the Galaxy. In an extrapolated geocentrism, Herschel came to the conclusion that the Milky Way encircled the Earth and that, like some of the nebulae in deep space, the Galaxy was the shape of a flattened disk. The conclusion was clear. The Earth, or at least the Sun about which it moved, was at or near the center of the Galaxy. Though Herschel's methodology was flawed, his model remained relatively unscathed for the next one hundred years, given the odd minor tweak.

Paradoxically, even though he was a galactocentrist, by dwelling on the very construction of the cosmos, Herschel struck out against anthropocentrism. A prophet of deep space astronomy, while others peered at the familiar local planets, Herschel was probing the depths of distance and the unidentified. While professional astronomers were tinkering with planetary distances in the Solar System, Herschel the amateur was charting star systems beyond the professional imagination. And while they were using the rough proportions of the speed of light to infatuate over the mechanics of the Jupiter system, Herschel realized he was gazing so far into deep space that he was looking millions of years into the past.

EINSTEIN'S INFLUENCE

The biological theory of evolution also had an effect on anthropocentrism. Though its immediate impact on the place of Man among terrestrial animals was quite profound, its influence on the question of alien life was more gradual. But as evolutionary theory threatened the image of Man due to his descent from lowly creatures, the even more all-embracing idea of pluralism promised a worldview in which Man was but one link in a cosmic chain of life that stretched far into deep space. By the beginning of Einstein's century this biological cosmos was a recognizable worldview. It informed the fictions of H. G. Wells, among others, and influenced the narratives of one of Wells's successors, Olaf Stapledon.

During the course of the twentieth century, the tension between anthropocentrism and pluralism grew. As Einstein's influence diffused into the scientific culture, the early decades saw radical changes in the astronomical worldview. Whereas it might still have been possible in 1903 to argue an anthropocentric cosmos, by 1930 evidence had all but annihilated that argument. It was as Einstein had predicted: an expanding Universe. But astronomers had also uncovered a cosmos of enormous dimensions. The Solar System seemed washed up on the shore of one Galaxy among millions. And a tipping point was passed after which the belief in other worlds became commonplace. The "assumption of mediocrity" or the "principle of terrestrial mediocrity" as Kant put it, became an integral belief in standard cosmologies.

The sciences began to merge as the century dawned. Both the physical and biological elements of the cosmos needed to be addressed to avoid future anthropocentric claims on our place in the Universe. Those elements of pluralism, the physical and the biological, became increasingly testable as the decades unfolded. Yet, the materialist or idealist assumptions, which lay beneath the interpretation of such scientific observations, remained.

LAST AND FIRST MEN

Einstein's new curved space-time was a gift to the writers of the world. His new sky became a theatre that unleashed the most incredible of concepts. A new stage was set, which included not just all of space, but all of time too. The astounding age of science fiction, those first few decades of the twentieth century, featured the pulp fiction of dizzying technologies, immense interstellar starships, and fleet and nimble faster-than-light vessels worming through space-time. The stories that at first were merely pulp would later prove hugely profound.

The evolution of mankind into a star-faring species was a concept explored in much detail in the work of Olaf Stapledon. Born in 1886 in the Liverpool area of England, Stapledon was a philosopher and an innovative and influential science fiction author. Awarded a PhD in philosophy from the University of Liverpool in 1925, he soon turned to fiction in the hope of presenting his ideas to a wider public.

For our potted history of alien fiction, Stapledon produced a key work in his 1930 novel *Last and First Men*, an anticipatory history of eighteen successive species of humanity. Stapledon's writings greatly influenced not only key players in our own story on alien fiction, such as Arthur C. Clarke and Stanisław Lem, but also figures as diverse as Jorge Luis Borges, Bertrand Russell, Virginia Woolf, and Winston Churchill.

In Stapledon was a writer equipped for the age of Einstein. *Last and First Men* synthesized astronomy and evolutionary biology. It conjured up cosmic myths suitable for a skeptical and scientifically cultured age. As American author and astrophysicist Gregory Benford comments in his introduction to *Last and First Men*, "Stapledon had studied the principal scientific discovery of the nineteenth century, the Darwin-Wallace idea of evolution and projected it onto the vast scale of our future, envisioning the progress of intelligence as another element in the natural scheme. In his hands pressures of the environment become the blunt facts of planetary evolution, the dynamics of worlds."

EVOLUTION AND EINSTEIN

Evolution and Einstein transformed the concept of the alien. And as biology and relativistic cosmology found their way into twentieth-century fiction, the alien became one of the enduring and universal motifs of the age. Consequently, an increasing number of people met these revolutionary ideas not through science, but as a text, inspiring emotional as well as intellectual reactions. Thus, ideas of life in the new Universe were embedded more deeply into the public psyche, and scientific ideas were creatively morphed into symbols of the human condition that were often an unconscious and therefore particularly valuable reflection of the assumptions and attitudes held by society. By virtue of its ability to project and dramatize, science fiction has been a particularly effective and, perhaps for many readers, the only means for generating concern and thought about the social, philosophical, and moral consequences of scientific progress.

In turn, as scientists are creatures of the culture in which they swim, alien contact stories of the late nineteenth and early twentieth centuries motivated a significant number of scientists. The idea of life in the Universe, and Man's place within it, was firmly fixed in the scientific as well as the popular imagination. The theory of evolution had given credence to the evolution of life on Earth, and to evolution in a cosmic setting. The new sciences of biology and cosmology inspired a wealth of fiction and provided a rationale for imagining what cosmic life might develop. From now on the idea of cosmic life became synonymous with the physical and mental characteristics of the alien. It provided a new rubric against which man himself could be measured.

H. G. WELLS IS SHAKESPEARE, OLAF STAPLEDON IS MILTON

As we have seen, the impact of H. G. Wells's *The War of The Worlds* was responsible for igniting the alien theme in science fiction, and in the public imagination. Wells created the modern nexus of the alien, armed with its potential for probing human evolution. Wells's books are, in their degree, myths; and Wells is a mythmaker. Once developed by Wells, the modern alien idea proved a potent motif for cultivating fictional explorations of the singularity or insignificance of humanity. During such explorations, the secondary question of the character of alien and interspecies interaction became an issue, which later affected the SETI science program. As English science fiction writer Brian Aldiss put it, "Wells is the Prospero of all the brave new worlds of the mind, and the Shakespeare of science fiction."

If H. G. Wells is the Shakespeare of science fiction, then Olaf Stapledon is its Milton. Stapledon used the science fiction genre to explore nothing less than the meaning of human existence in a cosmic setting. And his books attempt to imagine and project the prospect of physical contact with alien life. Witness Stapledon's account of the Martian invasion of Earth. It is typical in the way that Stapledon's humans and aliens alike cross the void between the stars in vessels unimaginable to moribund Earth dwellers. Viewed from the confines of a contemporary pedestrian existence, Stapledon's vessels journeyed the skies above in both architecture inconceivable and technology incredible. His novel *Last and*

First Men invented the UFO: "Early walkers noticed that the sky had an unaccountably greenish tinge, and that the climbing Sun, though free from cloud, was wan. Observers were presently surprised to find the green concentrate itself into a thousand tiny cloudlets, with clear blue between . . . though there was much that was cloud-like in their form and motion, there was also something definite about them, both in their features and behavior, which suggested life."

MYTH MAKER

In the preface to *Last and First Men*, Stapledon tells the reader that his story is an attempt "to see the human race in its cosmic setting, and to mold our hearts to entertain new values." In a telling evocation of the theories of evolution and relativity, he suggests that such attempts to extrapolate Man's evolutionary future "must take into account whatever contemporary science has to say about Man's own nature and his physical environment." Stapledon produced a fiction that incorporated the most recent ideas of cosmology and evolutionary biology, thus creating a new fusion of fact and fiction, a form of fable for a scientifically cultured twentieth century. In the words of Stapledon, the aim must not be just "to create aesthetically admirable fiction . . . but myth."

That Stapledon was well aware of the latest developments in astronomy is evident from one of his visions on the human future in space: "But in the fullness of time there would come a far more serious crisis. The Sun would continue to cool, and at last Man would no longer be able to live by means of solar radiation. It would become necessary to annihilate matter to supply the deficiency. The other planets might be used for this purpose, and possibly the Sun itself. Or, given the sustenance for so long a voyage, Man might boldly project his planet into the neighborhood of some younger star . . . He might explore and colonize all suitable worlds in every corner of the Galaxy, and organize himself a vast community of minded worlds."

Stapledon's imagined life in the Universe is one of emerging genetics. He imagined the future forms of Man in an alien setting. *Last and First Men* is a future history on a staggering scale. The "hero" of the book is not a man, but Mankind. The story embraces seventeen evolutionary mutations, from the present "fitfully-conscious" First Men, to the glorious godlike Eighteenth Men who reign on Neptune. It is a history that spans two thousand million years.

Eugenics was the dark side of Darwinism. The first and most infamous exponent of a genetic intervention in the human race was Darwin's cousin, Francis Galton. It was Galton who introduced the word eugenics into the lexicon. In *Last and First Men*, Stapledon thought long and hard about eugenic practices. One of the causes of the demise of the First Men, for instance, was their failure to realize a eugenics program: "In primitive times the intelligence and sanity of the race had been preserved by the inability of its unwholesome members to survive. When humanitarianism came into vogue, and the unsound were tended at public expense, this natural selection ceased. And since these unfortunates were incapable alike of prudence and of social responsibility, they procreated without restraint, and threatened to infect the whole species with their rottenness." So, human intelligence steadily declined, "And no one regretted it."

Later came the irresistible rise of the Third Men. With their rediscovery of eugenics, the Third Men focused their efforts on that most distinctive feature of Man, the mind. Seeking to "breed strictly for brain, for intelligent coordination of behavior," the peak of their engineering program was the Great Brains. The Great Brains first helped, then enslaved, and finally annihilated their creators. Ultimately, they turned their cool intellects upon themselves. They created a superior species, the Fifth Men. The Fifth Men were accomplished in art, science, and philosophy, perfectly proportioned of body and mind. They were able to travel mentally back through time to experience the whole of human existence. Indeed, the Fifth Men became the most perfect species ever to dwell on planet Earth.

Stapledon's fiction, then, emphasized the triviality of humanity in the face of a new and vast cosmos, which itself may harbor truths and meaning as yet unknown to an immature human race. His fiction on the question of intellectual contact with alien biologies had a lasting influence on the twentieth century to come. Working scientists, such as J. B. S. Haldane, physicist Fred Hoyle, and Carl Sagan, one of the founders of SETI in the early 1960s, were swayed by Stapledon, as was fellow British science fiction writer Arthur C. Clarke.

ALIENS IN THE SPACE AGE

"We take off into the cosmos, ready for anything: for solitude, for hardship, for exhaustion, death. Modesty forbids us to say so, but there are times when we think pretty well of ourselves. And yet, if we examine it more closely, our enthusiasm turns out to be all a sham."

—Stanisław Lem, *Solaris* (1898)

TESLA TO SETI

He died without much money to his name, but Nikola Tesla lived a life rich in color. Widely esteemed as one of the greatest engineers working in America, Tesla was born in 1856, an ethnic Serb in a village that is now part of Croatia. Variously an inventor, engineer, and businessman, Tesla's theories and patents in electromagnetism formed the basis of wireless communication and radio.

But Tesla was also at the forefront of alien contact. As early as 1896, he proposed that radio could be used as a form of contact with alien life. And while experimenting on atmospheric electricity using a Tesla coil receiver at his Knob Hill laboratory in Colorado, he reported observing repeated (very probably artificial) signals, which Tesla interpreted as being of extraterrestrial origin.

After the second world war, physicists Phillip Morrison and Giuseppe Cocconi identified the potential of the microwave region of the electromagnetic spectrum, proposing a set of initial targets for interstellar communication. A year later in 1960, astronomer Frank Drake from

Cornell University made the first modern SETI experiment. Christened Project Ozma after the Queen of Oz in L. Frank Baum's fantasy books, Drake employed a radio telescope at Green Bank in West Virginia. With this scope he examined the stars Tau Ceti and Epsilon Eridani close to the 1.42 GHz marker frequency, a region of the radio spectrum known as the "water hole" due to its propinquity to hydrogen and hydroxy radical spectral lines.

The first SETI conference took place at Green Bank in 1961, and funding programs for an actual scientific search of alien intelligence began in earnest. Scientists from the Soviet Union soon took a strong interest, and in the 1960s performed a series of searches in the hope of detecting powerful radio signals from space. Indeed, international cooperation in the field was heightened by the definitive book *Intelligent Life in the Universe* (1966), written by USSR astronomer Iosif Shklovskii and US astronomer Carl Sagan.

This book was fascinating. Not only did Sagan and Shklovskii provide a scientific introduction to the subject that was soon to be better known as astrobiology, but they also devoted chapters to alien contact. Though they stressed their ideas were speculative, Sagan and Shklovskii nevertheless argued that scholars should countenance the possibility that alien contact had occurred during recorded history. Further, they suggested that sub light-speed interstellar travel by a sophisticated alien civilization was a certainty, especially when considered against the developing pace of technological progress in the late 1960s.

Perhaps inspiring the wave of ancient astronaut books that proved popular in the 1970s, Sagan and Shklovskii became more daring. They maintained that recurring occurrences of alien visitation to Earth were credible, and that prescientific myths possibly describe contact with aliens. They cited the tale of Oannes, a fish-like being in several rational and independent ancient sources, attributed with introducing agriculture, art, and mathematics to the early Sumerians. They urged the global community of scholars to scrutinize ancient documents for further possible instances of paleocontact.

SOLARIS

As the scientific world was abuzz with the idea of contact, science fiction produced a skeptic to match the hype. Stanisław Lem's *Solaris* (1961) is ultimately about the problems of contact, and the shortcomings of any potential communication between Man and alien. In exploring and probing the ocean planet Solaris from an orbiting research station, the human scientists too are being probed. The planet is sentient and delves into the minds of the humans who are analyzing it. The ocean planet is able to manifest guilty secrets in human form, which each scientist is forced to personally confront.

Described in 1976 by fellow fiction writer Theodore Sturgeon as the most widely read science fiction author on the planet, Stanisław Lem is perhaps best known for *Solaris* in all its forms: the 1961 novel along with two cinematic interpretations, the 1972 adaptation by acclaimed Russian film director Andrei Tarkovsky and the 2002 remake by US director Steven Soderbergh. Despite its undoubted popularity, until 2012 there had been only one English version of the text, translated in a rather stilted manner from French. According to Lem, this has led to inaccurate interpretations of Freudian messages in the text: "One American reviewer made a fatal mistake in that he was unaware of the fact that the idioms of the Polish original are different—hence they do not allow such [Freudian] conclusions."

Freudian nuance notwithstanding, the story arc of *Solaris* is apparent enough. The tale unfolds onboard an orbiting research station high above the ocean planet of Solaris. The book's narrator, Dr. Kris Kelvin, a psychologist, gets to Solaris after the long journey from Earth. His brief: to gauge the viability of the continued study of the planet, as precious little progress had been reported to Earth. Kelvin finds a cosmic Mary Rose. The station is deserted, save an elusive African woman and a Dr. Snow, who behaves peculiarly on meeting, probing Kelvin's identity and even questioning his existence.

As the tale unfolds it becomes clear to Kelvin that the crew, himself included, are experiencing the materialization of physical human simulacra, or "Phi-creatures." Through these creatures, each scientist is, in turn, confronted with their most painful and repressed thoughts and memories. Witness the "giant Negress," Dr. Gibarian's visitor, who twice appears to Kelvin, and seems to be unaware of the other humans she meets, or simply chooses to ignore them. His own Phi-creature, Rheya, is a simulacrum of his dead wife, whose suicide a decade earlier left Kelvin wracked with enduring guilt.

Lem uses techniques to enhance the threatening and alien feel of the Phi-creatures. The scientists on board the orbiting station probe the makeup of the creatures. Kelvin, along with Drs. Sartorius and Snow, attempts an analysis and understanding of Solaris and its associated phenomena. They carry out tests. Kelvin takes and investigates blood samples from the Rheya simulacrum, finding the Phi-creatures to be neutrino-based. The discovery eventually leads to the humans making an antineutrino gun capable of destroying the Phi-creatures, as all other attempts to terminate them met with failure. Though human in appearance, they seem indestructible. Kelvin previously tried "removing" the Rheya simulacrum in one of the escape pods. On her return, and upon realizing she is not the real Rheya, the simulacrum unsuccessfully tries suicide by drinking liquid oxygen.

During his stay on the station and cut off in his cabin with only the Rheya simulacrum for company, Kelvin attempts to understand the ocean planet. He delves into the research station archives, a library holding hundreds of volumes, which represent a century of research into Solaristics. But more is less for Kelvin. As he ploughs through the archives, the more he realizes that Terran science will never understand the alien complexities of the ocean planet and its phenomena. Likewise, the crew continue their attempts at contact with the ocean through x-ray stimulation linked to Kelvin's brain activity. All comes to naught and Kelvin is left questioning the very idea of contact and the nature of life in the Universe, concluding that the ocean planet is a flawed God-like sentience that cannot be understood by human science.

THE PROBLEMS WITH CONTACT

The moral of Lem's message is clear. The stars are not for us. No matter how much we desire contact and communication with an alien civilization, it may never be achieved. Terran science may not decipher something so alien from the human, making a symbiotic communication between alien civilizations nearly impossible. Indeed, Lem also explored the alien contact theme in two other novels, *Eden* (1959), in which there is face-to-face contact with aliens but a barrier to translation, and *His Master's Voice* (1968), where, as with *Solaris*, the barrier of translation is further explored by taking away the possibility of face-to-face contact.

Lem's *Solaris* is a book about the culture of science and its impact on contact. In particular, it focuses on the ideology of scientific culture which, in the context of Lem's narrative, revolves around the science of Solaristics. The surface of the Solaris ocean is composed of a colloidal substance that can morph into many shapes, some of which ("mimoids") are simulacra of terrestrial objects. The ocean planet has so long been a subject of study for Terran scientists that Solaristics evolved into an institution. And so the story centers on the history and controversies of this alien science. The nature of the planet itself, as an example of alien life, and the empirical data gleaned by the research station also feed into Solaristic studies.

The other obsession in *Solaris* is the question of coding. As we have said, the Phi-creatures (also called Visitors, or "polytheres") are a kind of neural projection, materializing from the brains of the human scientists onboard the research station. Their origin lies in the most enduring imprints of memory, those that are well rooted, but of which no individual imprint can be singled out. And any attempt to analyze the intent of the occurrences of the Phi-creatures is doomed to failure, blocked by the anthropomorphism of the scientists who "own" the memories. As Lem would have it, and in Freudian terms, the Solaris ocean has made flesh the forces of the id. And the blocking of the "owners" represents the hypothesized Freudian mechanisms that function between the id and the ego.

Solaris is a psychic ocean. In the same way that Terran scientists are able to read genetic information, coded into the sequences and structures of the DNA located on chromosomes, the ocean planet is able to understand the physico-chemical psychic processes of the brain. In the final analysis, the logic goes, any human explanation of learning and memory ultimately involves some amalgamation of neurons, RNA, and proteins, the chemicals of life through which hereditary information is stored and reused. The

Solaris ocean is able to understand and retrieve such stored imprints and make flesh the psychic tumors of each scientist on the research station.

THE LIMITATIONS OF SCIENCE

In this way, Lem uses *Solaris* to focus on two limitations of the culture of science: its anthropocentricism, and its institutionalized nature. These two limitations of science are shown by Lem to produce a damaging mysticism. Kelvin arrives on Solaris with all good intentions, remembering "that thrill of wonder which had so often gripped me, and which I had felt as a schoolboy on learning of the existence of Solaris for the first time." But Lem's story traces Kelvin's decline into mysticism. In the end, the despair Kelvin feels is paralleled by the evaporation of his confidence in the lacking explanations of mainstream science. To make matters worse, there is the Rheya simulacrum. The futile relationship Kelvin has with the reincarnated Rheya serves to focus on the subjective side of science. His emotions prevail over his scientific reason, which accentuates the latter's brittleness. It is only Rheya's suicide that liberates him once more to focus on the question of Solaris.

Kelvin's is a journey into the dark heart of science. As he delves into the history of Solaristics, he becomes increasingly disillusioned about the ability of science to explain alien phenomena. Lem carefully chooses the character of Kelvin. Through him as narrator, the reader gets a strong sense of the self-defeating journey upon which Kelvin, and through him not just Solaristics but all "alien" science, is embarked. The field was once vital but now all creativity and originality are gone. All that's left are data collation and the shoveling of new facts into a degenerated and machine-like worldview.

PARADIGM SHIFT

Lem engages the alien life debate from the perspective of paradigm. At the time of the writing of *Solaris*, American historian and philosopher of science Thomas Kuhn was making his mark with his analysis of the progress of scientific knowledge. In two books, *The Copernican Revolution* (1957) and *The Structure of Scientific Revolutions* (1962), Kuhn developed the idea that science undergoes periodic "paradigm shifts." When one worldview, or paradigm, gives way to another, all scientific challenges are met from within the boundaries of this new framework. For example, Aristotle's geocentric cosmos into Copernicus's heliocentric Universe. And importantly, the notion of scientific truth, at any time, cannot be established solely by objective criteria, but is defined by a consensus of a scientific community within that paradigm.

In the terms defined by Thomas Kuhn, Solaristics needs a new paradigm. The current contested paradigm of Solaristics flies in the face of the facts implied by the alien phenomena. But the alternative paradigm the data implies is simply out of reach. The paradigms are incommensurable; they are competing accounts of reality, which cannot be coherently reconciled. The Terran comprehension of the cosmos can never be truly objective. Indeed, it is Gibarian, the most skeptical but optimistic of the scientists, who came closest to providing this new perspective in his contact with the ocean.

It's a kind of geocentrism in the very nature of science itself that is a barrier to insight. Though Kelvin understands these limitations of science, it is Snow who has to point out that the geocentrism of science is the root cause. As Snow says, "We don't want to conquer the cosmos, we simply want to extend the boundaries of Earth to the frontiers of the cosmos. We are only seeking Man. We have no need of other worlds. A single world, our own, suffices us; but we can't accept it for what it is. We are searching for an ideal image of our own world: we go in quest of a planet, a civilization superior to our own but developed on the basis of a prototype of our primeval past. At the same time, there is something inside us which we don't like to face up to, from which we try to protect ourselves, but which nevertheless remains, since we don't leave Earth in a state of primal innocence. We arrive here as we are in reality, and when the page is turned and that reality is revealed to us—that part of our reality which we would prefer to pass over in silence—then we don't like it anymore."

Terrestrial science is not fit for extraterrestrial exploration. Snow suggests that the inherent limitations of Solaristics render it unfit for the mission of alien contact and communication. Snow will not replace science with the divine and he rejects the proposals of Kelvin to substitute mysticism for methodology in future scientific training courses. Finally, though the other scientists elect to leave the orbiting research station, Kelvin decides to stay. As Kelvin cites his reasons, we catch another glimpse of the contradictions at the heart of the research. Kelvin wishes to remain "in the faith that the time of cruel miracles was not past," referring to the "miracles" of contact with the Solaris ocean and the revisit of Rheya, though he fully recognizes his desperate and alien situation.

Lem is the alien contact skeptic, the lone voice in the wilderness of deep space. *Solaris* is his manifesto on the limitations of science. It is a tale in which science is limited not only as a methodology, but also as a faith. And the provenance of the limits of science is rooted in its geocentrism. But *Solaris* is also a despairing novel. It suggests no plausible alternatives. Written at the time of the instauration of SETI, Lem portrays scientists confronted with extraterrestrial phenomena that exceed the limitations of their terrestrial science. Science may have been developed into theoretical quantitative Solaristics, but it cannot account for actual experience of alien contact and communication.

THE FUTURE ALIEN

Maybe tomorrow, or maybe a decade or century from now, we will make the most shattering discovery of all time: the discovery of a thriving alien civilization. As the twenty-first century dawned, we had been imagining alien life for almost two and a half millennia. But as space agencies prepared to build flotillas of space telescopes to search for life in this unearthly Universe, the crucial questions remain unanswered, as Lem's work showed.

Today, the plenitude principle abounds. It was the philosopher Arthur Lovejoy who first traced a history of the belief in the abundance of creation. Lovejoy identified its philosophical origin with the Greeks such as Epicurus and his follower Lucretius, the Roman. Since those early days, though both idealists and materialists used the idea of plenitude, a Copernican take on the principle led to the belief in an infinity of alien worlds.

We live in a great age of discovery. An age about which writers and scholars could only have dreamed. Astronomers hunting for potentially life-bearing terrestrial planets around Sun-like stars estimate there may be tens of billions in our Galaxy alone. A European team of scientists reported that perhaps 40 percent of the estimated 160 billion red dwarfs in the Milky Way have a "super-Earth" orbiting in a habitable zone that would allow water to flow freely on its surface.

In a very real sense, the Copernican Revolution has been reborn. The American space observatory Kepler, launched in 2009 to find Earth-like planets orbiting other stars, took off four hundred years after Galileo's first use of the telescope (and is of course named after the first great Copernican theorist, Johannes Kepler). Based on Kepler's early findings, Seth Shostak, senior astronomer at the SETI institute, estimated that within a thousand light-years of Earth, there are at least 30,000 habitable planets. And based on the same findings, the Kepler team projected that there are at least 50 billion planets in the Milky Way, of which at least 500 million are in the habitable zone. NASA's Jet Propulsion Laboratory (JPL) was of a similar opinion. JPL reported an expectation of two billion "Earth analogues" in our Galaxy and noted around 50 billion other Galaxies potentially bearing around one sextillion Earth analogue planets.

Over the last two and a half thousand years, a stunning array of scholars, scientists, philosophers, filmmakers, and writers have devoted their energies to imagining life beyond this Earth. Their task has been to reduce the gap between the new worlds uncovered by science and exploration, and the fantastic strange worlds of the imagination. Their huge contribution has been important not only in the way that the fictional imagination has helped us visualize the unknown, but also for the way in which it has helped us define our place in a changing cosmos.

A mode of thinking has emerged. One in which the science and culture of astrobiology have been synthesized. In the rich evolution of the alien life debate, science fictional visions have influenced issues and dialogues in astrobiology, and in turn popular culture has been inspired by discovery and invention. The history of astrobiology has hinted at the revolutionary effects on human science, society, and culture that knowledge of another civilization will bring. If we may be so bold as to suggest that humanity is at least a way in which the cosmos can know itself, what more is out there to be discovered?

PART II

IT'S LIFE, JIM, BUT NOT AS WE KNOW IT

WHAT DO XENOMORPHS FROM *ALIEN* AND NA'VI FROM *AVATAR* HAVE IN COMMON?

"Nowhere in space will we rest our eyes upon the familiar shapes of trees and plants, or any of the animals that share our world. Whatsoever life we meet will be as strange and alien as the nightmare creatures of the ocean abyss, or of the insect empire whose horrors are normally hidden from us by their microscopic scale."
—Arthur C. Clarke, *Profiles of the Future* (2013)

INVENTING THE ALIEN

How do writers and movie makers imagine the unimaginable when it comes to alien life? It's pretty tricky, imagining the unknown. Authors and film directors have come up with some very different ways of representing the unimaginable. The creative minds of Hollywood and beyond have populated cinema and television with a veritable zoo of extraterrestrials, from the dread-inducing xenomorph of the Alien franchise to the more benign Na'vi from *Avatar*.

Let's take a brief magical mystery tour around some of the more famous aliens in recent movie history. Back in 1977, a seminal year for science fiction cinema, the slender humanoid aliens from *Close Encounters of the Third Kind* had big heads and small bodies and were roughly about the height of a human child but with a much taller leader with very long limbs. In the same year, *Star Wars* brought us, among other things, the Wookiee. These long-lived, shaggy giants had volatile tempers but were also clearly sophisticated and intelligent, as they knew how to engineer a starship and navigate it through hyperspace. By the close of the decade, we had the very alien creature showcased in *Alien*, the first movie in what was to become a franchise. This extremely predatory extraterrestrial is sometimes known as the xenomorph. But the word xenomorph, which means "alien form" from the Greek *xeno*, meaning "other" or "strange" and *morph*, meaning shape, was first used by the *Aliens* character Lieutenant Gorman with reference to generic extraterrestrial life. The word was mistakenly assumed by some fans to mean that specific creature in *Aliens*. Nonetheless, what a creature it was. And how influential. Described as a perfect organism averaging seven to eight feet tall, with Queens twice as large, these beings had multiple life cycles that needed a host organism to reproduce. The xenomorph was immensely fast, strong, and intelligent, with an armor-like carapace, a bladed tail, and acid for blood.

ASOGIANS TO GROOT

The year 1982 brought us *E.T. the Extra-Terrestrial*. The Asogians, the alien race to which ET belongs, are short, squat, brown-skinned sentient creatures with bulbous torsos, long arms, and curiously extendable necks, somewhat similar to turtles and snails. ET's lot are a touchy-feely bunch. Their hearts glow to show emotion, they're able to heal others with the touch of their fingers, and they also appear to have psychic abilities such as telepathy and telekinesis, a quite common trope in sci-fi to suggest an advanced intelligence. Yautja, the extraterrestrial creature in the 1987 movie *Predator* was of the xenomorph variety: a highly intelligent and advanced species, they travel the Galaxy in search of dangerous prey to hunt for sport. The 1990s then brought us the "Harvester" alien in the 1996 movie *Independence Day*, telepathic creatures with a hive mind capable of controlling other species by touching them with their tentacles, in some kind of weird mix between ET and the xenomorph.

In the latest generation of movie aliens, we've had *Avatar*'s omnivorous hunter-gatherer humanoid aliens known as Na'vi, *District 9*'s opportunistic bipedal insectoid carnivores known as Poleepkwa, and *Guardians of the Galaxy*'s large humanoids, which resemble trees and are able to control their bodies to expand and contort themselves into different shapes and forms. These last shapeshifter aliens are known as Flora Colossus, or maybe better known by "I am Groot."

ALIEN CLOUDS AND OCEANS

Some writers' depictions of extraterrestrials in literature are a little braver, maybe because the writers don't have the drawback of having to present their conceptual aliens on the silver screen. In his first novel, *The Black Cloud*, published in 1957, British physicist Fred Hoyle explored the idea of a disembodied alien intelligence. The story begins with the discovery of the interstellar cloud of matter and its implications for science in coming to terms with the form and nature of alien life.

Another challenging portrayal in alien fiction, and one which *did* make the movies, was Lem's *Solaris*. This 1961 book was made into two movies, one in 1972 by acclaimed Russian film director Andrei Tarkovsky and one in 2002 by US director, Steven Soderbergh. The story is about the ocean planet of Solaris. The alien is the ocean, is the planet. Indeed, Solaris is a *psychic* ocean. And as the orbiting human scientists try to "read" Solaris, Solaris is reading *them*.

WHAT IS LIFE?

What do these few introductory alien portrayals have in common? Well, for one thing, their scientific starting point. When dreaming up an alien species, writers have to wonder about what we all mean by "life." How can we tell if something is alive? A PlayStation 5 has a number of hidden but moving parts; is it alive? Luckily for writers and moviemakers, planet Earth is home to a great variety of living things, from microscopic bacteria to trees so tall their tops can't be seen from the ground, from the smallest insect to the most enormous whale. Many creatures can be scaled up to make itself an alien and, no matter how simple or complex, all living things show seven characteristics of life.

LIFE . . .

. . . IS MADE UP OF CELLS

Living plants and animals are organized from one or more cells, as cells are the building blocks of life on Earth, and maybe the building blocks of life elsewhere. Unicellular organisms such as bacteria consist of just a single cell but others such as humanoid aliens are multicellular. We include plants in our definition because, of course, some stories revolve around an alien plant life-form such as *Invasion of the Body Snatchers* and *The Day of the Triffids*.

. . . GROWS

Some stories might allow for the fact that organisms get bigger, growing in size from offspring to adult. This organized growth is not random, it is another feature definition of life. Creatures usually increase in size in all their parts equally rather than simply growing willy-nilly. Such inconsistent growth might leave a being's arms and legs puny while they grew the fattest of heads. Having said that, this situation adequately describes the invading Martians in H. G. Wells's *The War of the Worlds* (but Wells was trying to make a point that the technologically superior Martians had atrophied limbs and oversized heads because of their superiority).

. . . USES ENERGY

All living creatures use energy and transform it in some way. We humans do this with food, converting the constituents in the food into energy that our cells can use. There's also the question of having to decompose the waste organic matter, also known as pooping, though this is a science question rarely broached in sci-fi movies. Having said that, in a July 2017 issue of *New Scientist* magazine, it was reported that if we're going to look for signs of alien life on other planets, we'll improve our odds if we expand that search for the trail those organisms might have left behind. In other words, perhaps we should be looking for alien poop.

. . . RESPONDS TO SURROUNDINGS

All organisms respond to their environment. Even tiny bacteria respond to chemicals in their surroundings. When multicellular organisms respond, it can involve a number of complicated senses. But responses can be simple, too, like the leaves of a plant turning toward the Sun. Let's take the example of the British sci-fi classic *The Day of the Triffids*. A novel in 1951 and a movie in 1962, *The Day of the Triffids* is about an aggressive species of plant that starts killing people. The triffids respond to their environment by directing their stings at humans' heads. But those humans who wear helmets often escape unscathed. And yet the triffids continue to try to sting their opponents even if they are wearing protective gear, as it's a mechanical response stimulated by external shifts in their environment.

. . . REPRODUCES

Plants and animals produce offspring. If they make copies of themselves from just a single parent, it's known as asexual reproduction, and if they make copies from two parents, it's known as sexual reproduction. The chemical that does the job of copying is called DNA, of course, but more of that later. Consider reproduction in the Alien franchise. The so-called xenomorph life cycle consists of five main stages: the Ovomorph, the Facehugger, the Chestburster, the adult, and finally the Queen. Writers partly base this cycle on certain parasitoid insects found on Earth such as the wasps of the *Chalicidoidea* and *Ichneumonoidea* families which lay their eggs on live prey that are then consumed by the hatching larvae.

. . . EVOLVES!

Yes, life on Earth, and we assume life beyond, changes over a period of time. Life evolves, in short. But given that evolution is the change in the characteristics of a species over several generations and relies on the process of natural selection, it's hardly the kind of dramatic thing that can be easily filmed. But Arthur C. Clarke and American director Stanley Kubrick still tried making a film about evolution. Their 1968 movie *2001: A Space Odyssey* identifies three stages in the evolution of humans: ape, modern humans, and, ultimately, super-humans. *2001: A Space Odyssey* was called a "scientific definition of God" by Kubrick. Even though the theory of evolution proposed by Darwin was all about slow, inexorable change, Kubrick and Clarke tell a tale of human history by coupling the usual driving force of evolution, long periods of steady change, with the episodic guiding hand of superior alien beings. *2001* is a story of the effective creation and resurrection of humanity.

LIFE-GIVER

There are a number of unwritten laws about alien life that often don't make the movie cut but are there in the background. For example, for all life here on Earth, the Sun is the most important object in our planet's history. Not just because our local star's life-giving powers, as nearly all organisms are dependent on energy from the Sun, but also because the actual distance of our home from the central fire is vital for the development of life on Earth.

Plants capture just a small percentage of the Sun's energy that reaches them. They convert Sunlight into food energy that is subsequently stored. This light conversion process of plants is, of course, known as photosynthesis, which just means "collecting light." On the face of it, it doesn't seem like the kind of topic you might write a science fiction story about. But British sci-fi writer Adam Roberts wrote an ingenious tale in 2012 called *By Light Alone*. In a world where we humans have been genetically engineered so that we can photosynthesize Sunlight with our hair, hunger is a thing of the past, and food an indulgence. The poor grow their hair, the rich affect baldness and flaunt their wealth by still eating. And that's just the start of the story!

Photosynthesis is just one small part of food webs, or food chains, whereby the source of all clean "green" energy for the ecosystems and communities of living organisms on Earth is passed from one organism to another. If life exists elsewhere, we can assume that creatures of another world will evolve and become part of similar ecosystems, perhaps, in orbit about other stars like our Sun. On those alien worlds, an ecosystem may begin with vegetarian animals that eat plants and absorb energy. The alien plant-eaters are then munched by alien meat-eaters, but the meat-eaters also prey upon each other for food energy. In this way, if a writer or moviemaker creates what scientists call a food web, then the fiction will be all the more realistic.

Finally, consider the question of the chemical elements. The stars of our Galaxy and of all other Galaxies in the cosmos are, simply put, fiery balls of burning gas. A star spends its life burning very basic elements such as hydrogen and helium to produce new and heavier elements such as oxygen, nitrogen, carbon, and so on. Such elements would not just be the very basic building blocks of the chemistry of life of all living organisms on our planet, but, as far as we can tell, all other alien worlds in creation.

LOST IN SPACE: ARE SILICON SPIDERS A THING?

"We are stardust/Billion year old carbon/We are golden."

—Joni Mitchell, *Woodstock* (1970)

"The atoms of our bodies are traceable to stars that manufactured them in their cores and exploded these enriched ingredients across our Galaxy, billions of years ago. For this reason, we are biologically connected to every other living thing in the world. We are chemically connected to all molecules on Earth. And we are atomically connected to all atoms in the Universe. We are not figuratively, but literally stardust."

—Neil deGrasse Tyson, *Cosmos: A SpaceTime Odyssey* (2014)

LOST IN SPACE

Lost in Space goes way back. It started as an American TV series on CBS and ran between 1965 and 1968, so it predates *Star Trek*. The idea for *Lost in Space* was inspired by *Swiss Family Robinson*, a novel written in 1812, along with a comic entitled *Space Family Robinson*. The story features the adventures of the titular Robinsons, a typical pioneering American family of space colonists who struggle to survive while lost in the depths of space. (The original 1812 story is about a Swiss family of immigrants whose ship en route to Australia goes off course and is shipwrecked in the East Indies. What's good enough for Earth is good enough for any other planet.)

Indeed, that original story has seen many versions, including a 2018 Netflix series adaptation and a 1998 movie adaptation, both based on the 1965 TV series. The 1998 film version contains the interesting alien life-form of space spiders. Okay, one immediate problem. The creatures are called "spiders" for dramatic effect but, since they only have six (rather than eight) legs, they're technically considered insects instead of arachnids. That said, it's not this inaccuracy that concerns us here. These alien critters can eat through metal and possess a jagged-toothed maw. They have a shell made of Admantium, which is of interest, as Admantium is clearly a fictional metal alloy (the Latin suffix -ium means "metal-like"). The movie's use of Admantium may be a play on Adamantium, a nearly indestructible metal alloy in the Marvel Universe. These critters also have communication skills like terrestrial bees and are known for being attracted to light and heat. Not only do they bite and eat humans, but a bitten person can slowly morph into a spider/human hybrid, which is pretty cool, though apparently unwelcome.

On analysis, it seems that these space spiders are the first life-forms known to human science that are *silicon*-based, not carbon-based. They even make a creepy metallic noise as they crawl about the passages of spaceships. Now, sci-fi is often suggesting hypothetical types of biochemistry, alternate forms of biochemistry that are agreed to be scientifically viable but not proven to exist at this time. And these biochemistries revolve around some chemical other than DNA to do the replication, some solvent other than water to be the matrix of life, and some chemical element other than carbon upon which life itself is based. So, is silicon-based life a possibility? To answer that question, let's take a look at how the cosmos makes its chemicals.

WE ARE STARDUST

We've seen that stars power the Universe as bringers of energy and light. They are the building blocks from which the entire cosmos was created. Without our own star there'd be no light, no life, and no Earth as we know it, so it's fair to assume that all alien worlds get their light and life from stars like the Sun. We spoke before of the Sun as a huge ball of burning hydrogen gas. About a quarter of its mass is helium gas. Why? Because stars like the Sun burn about four million tons of gas every second. That's as much energy as seven trillion nuclear explosions every second. At the very center of stars like the Sun, with temperatures over 10 million degrees, it's hot enough to fuse hydrogen gas into helium. That's how stars burn and fuse heavier elements.

The stars around which alien worlds orbit have compositions depending on how old they are. For example, more highly evolved stars have a greater amount of gases heavier than hydrogen and helium. Most of the material Universe is locked up inside stars. About three-quarters of this material is hydrogen, and one-quarter helium. But a tiny 2 percent of the cosmos is made of what astrophysicists call the "heavy" elements (some chemists laugh at this, as in chemistry the phrase "heavy elements" means something quite different). In astronomy, heavy elements are all the basic chemical elements that are more complex than hydrogen and helium.

Most stars, like the Sun, live for a long time. The Sun spends ten billion years fusing hydrogen into helium and is currently around five billion years old. Other more massive stars have quite short lives, some in terms of mere millions rather than billions of years. But all stars evolve, of course, and when they do, the products of their fusion (the elements which they've essentially created) are recycled back out into space. And *this* means that the elements made by star fusion are free to be used in the creation of alien planets and alien life.

Neil deGrasse Tyson, and Joni Mitchell and Carl Sagan before him, is famous for saying we humans are made of stardust. Witness the iron in your blood, the calcium in your teeth, and the carbon from which your skin is made. All three elements (iron, calcium, and carbon) are cooked up inside stars by fusion and then recycled back into space to make new stars, planets, and people.

CARBON IS COSMIC

Carbon is cosmic. Not only does it form the basis of life on Earth, it's also to be found out in deep space as a result of stellar fusion. Carbon is the backbone of biology on Earth because of its very nature. It easily bonds with life's other main elements like hydrogen, oxygen, and nitrogen. And carbon is also light and small, making it an ideal element for making the longer and more complex chemicals of life, such as proteins and DNA.

What makes carbon so cosmic? Think about its variety. It makes one of the *softest* known substances in graphite, and also one of the *hardest* known substances in diamond. In all, carbon is known to form ten million different chemicals, which is around 70 percent of all chemicals on Earth. In particular, carbon is made in the cores of giant and supergiant stars. It's then scattered into space as dust in supernova explosions. Some stars use carbon as a catalyst for their fusion reactions as they burn. And a number of complex carbon chemicals, including sugar, have been found in space.

WHAT ABOUT SILICON?

Carbon combines with hydrogen, oxygen, and nitrogen in making life's complex chemicals. This chemistry gives carbon the ability to form the long "hydrocarbon chains" of life. As silicon is in the same group of the periodic table of elements as carbon, writers and moviemakers like to suggest that, somewhere out in space, nature uses silicon rather than carbon. But most scientists believe that alien life would also be carbon-based. Critics call this carbon chauvinism, being narrow-minded about carbon alone. Non-carbon creatures may evolve in an environment that is not carbon-rich, they say.

But Earth has more silicon than carbon. The mechanisms of life, like breathing and photosynthesis, don't seem to work without carbon. They're carbon-*based*. But maybe life on Earth is quirky and life on alien worlds is very different. Yet, nature is believed to behave similarly throughout the cosmos. And it's worth remembering that while there's more silicon on Earth than carbon, carbon is still what nature chooses to use.

That's not to say silicon is impossible. Indeed, it's hard to imagine anything *more* likely than silicon, given that it's closest to carbon in terms of chemistry. But there are plenty of problems with this idea. Whereas carbon-oxygen bonds can be created and broken again (this goes on in our bodies all the time), the same isn't true for silicon. And that would limit silicon's life-like chemistry.

Maybe you could have silicon-based life-forms in the sense that they pass on information. Perhaps deep below the surface of an alien world that is a very hot, hydrogen-rich, and oxygen-poor environment, you could have a complex silicon chemistry of life. But with regard to silicon-based space spiders eating humans, they're more likely to eat our rock collections.

WHY DO THE HEPTAPODS IN *ARRIVAL* LOOK LIKE OCTOPUSES?

"A wonderful area for speculative academic work is the unknowable. These days religious subjects are in disfavor, but there are still plenty of good topics. The nature of consciousness, the workings of the brain, the origin of aggression, the origin of language, the origin of life on Earth, SETI, and life on other worlds . . . this is all great stuff. Wonderful stuff. You can argue it interminably. But it can't be contradicted, because nobody knows the answer to any of these topics."
                                          —Michael Crichton, *Profiles of the Future* (2013)

ARRIVAL

There have been some pretty unbelievable alien races in science fiction history. The Vogons from *The Hitchhiker's Guide to the Galaxy* were green-skinned, slug-like, bulky, vaguely humanoid, and bureaucratic creatures who wrote poetry and demolished the Earth. But, of course, *The Hitchhiker's Guide to the Galaxy* is that rare beast in sci-fi—a satire. The Vorlons in *Babylon 5* are also perhaps too unbelievable. An ancient and technologically advanced race, the Vorlons are super-psychic energy beings who genetically manipulated most of the younger races with daddy issues. When interacting with other races they wear complex encounter suits to completely hide their true appearance—a blatantly obvious budget-saving convenience. And yet one character who met the Vorlons claimed to have seen nothing at all, while another saw a very bright ball of energy. The 1953 movie *Robot Monster* features the monstrous Ro-Man, an alien that looks remarkably like a gorilla in a diving helmet, who has destroyed all but six people on Earth. Ro-Man spends most of the movie trying to finish off the human race but, in a reverse Captain Kirk kind of way, complicates matters when he falls for the young woman in the human group.

The aliens in the 2016 movie *Arrival* are a lot better thought-out. Let's remind ourselves of *Arrival's* plot. The film is based on Ted Chiang's award-winning 1998 science fiction tale *Story of Your Life*. As with the movie, *Story of Your Life* pictures Earth's contact with aliens in the form of heptapods who communicate in a cryptic language. Actress Amy Adams plays a linguist who tries to communicate with these creatures who have "invaded" Earth. In terms of pure CGI, the look of the alien heptapods is a tribute to how masterful the movies have become in representing alien life-forms. Their nuanced, fluid movements in Earth's denser atmosphere are elegant in their execution and, unlike many aliens before them, the hexapods seem credible—even *real*.

And there's a good reason for that credibility, one which goes back half a billion years. Maybe what makes the heptapods most credible is that they aren't all that alien to start with. Check out their body shape, or their tentacles, or their skill of squirting a kind of ink. In short, *Arrival's* heptapods bear an uncanny resemblance to Earth's most "alien" intelligent life: cephalopods. Like the octopus, the heptapod's huge eyes, its outreaching, slinking tentacles, and its boneless body make for a perfect extraterrestrial. They seem alien as they are very much "other," with little empathy or common ground, so why do the heptapods resonate so well with movie audiences, and how on Earth is the source of their credibility half a billion years old?

LIFE EXPLOSION

Ever since the beginning of Earth's history, life has evolved in leaps. For many millions of years, life crept along with very little change happening. For example, scientists think that unicellular life reigned supreme here on our planet for more than two billion years before the evolution and spread of multicellularity. Then, in sudden bursts of feverish activity, huge changes occurred. Of course, Earth's history stretches and creeps over billions of years, so when we say these leaps were sudden, we mean that they happened over mere *millions* of years. Compared to billions, millions really are quite quick.

Here's where evolution played its "half a billion years ago" trick. During the so-called Cambrian explosion, around 530 million years ago, nature had one of its evolutionary leaps. Fossil evidence in the rocks shows that before about 580 million years ago, most creatures were quite simple. But over the 70 or 80 million years of the Cambrian explosion, the variety of life began to rapidly change to resemble that of today. Most organisms' phyla, or body plans, appeared during the Cambrian explosion, along with a major diversification of creatures. Starting to get the hint?

In the study of life on Earth, a phylum is a group of organisms based around a body plan. The best-known animal phyla are: chordata, the phylum to which we humans belong, sponges, jellyfish, flatworms, roundworms, ringworms, echinoderms, arthropods, and mollusks. And it's the cephalopod mollusks, such as octopuses, which are among the most neurologically advanced of all invertebrates. Though there are 35 animal phyla in total, the 9 listed above include over 96 percent of all animal species on Earth. It was during the Cambrian explosion that many phyla first appeared. Since then, many species of animal have merely evolved upon the body plan that started with the Cambrian explosion. It's easy to imagine a similar explosion on an alien world that resulted in heptapod-like creatures evolving from that basic body plan ever since.

SCIENCE FICTION OCTOPOID

Given the alien nature of cephalopods like octopuses, it's hardly surprising that they have inspired the creative minds of moviemakers and writers. The pedigree is longer than you might think. In the books of nineteenth-century French astronomer and author Camille Flammarion, *Real and Imaginary Worlds* (1864) and *Lumen* (1887), Flammarion describes a range of exotic species, including a planet full of perpetually swimming, tentacled, seal-like beasts. Even H. G. Wells couldn't resist. The invading Martians in his 1898 novel *The War of the Worlds* had "Gorgon groups of tentacles."

Ever since those early days, octopoid creatures have been sci-fi fan favorites. So why do octopuses work so well in *Arrival*? The octopus is an excellent model for alien life. For one thing, Earth's oceans are like an alien world. A total of twelve humans have walked on the Moon, all under NASA's Apollo program. But to date, only five humans have been to the bottom of the Mariana Trench. The oceans also have an atmosphere we can't breathe, so nature gives birth to bizarre creatures beyond our land-based imaginations. Cephalopods are so very different from our intelligent mammalian cousins, such as chimps, yet octopuses nonetheless show amazing intelligence.

Octopuses use tools. For example, soft-sediment dwelling octopuses cart about coconut shell halves, using them as a shelter only when needed. Octopuses play, such as aiming water jets for the sheer fun of it. Octopuses solve puzzles and problems. An academic paper way back in 1990 showed that Octopus vulgaris Lamarck was able to open transparent glass jars closed with a plastic plug and containing a live crab. The creatures removed the plug and seized the crab in one swift attack. And the number of unsuccessful attacks decreased over a series of trials. Some scientists believe that octopuses may even engage in battle with improvised weapons. A study of the common Sydney octopus in Australia's Jervis Bay found that, when crowded, as octopuses are usually solitary creatures, they had taken to hurling objects at one another such as shells and seaweed and blasting them through the water with high pressure.

All this evidence confirms what scientists believe: octopuses are the only invertebrate animals that show a level of thinking usually associated with consciousness. Any reader who considers themselves a budding aquarist and who has tried to keep such critters in captivity will know not only how smart octopuses are but also just how much time and energy goes into keeping them happy and healthy.

MYSTERIOUS INTELLECT

So, there we have it—plenty of reasons why octopuses should be the basis for the heptapods in *Arrival*. Finally, consider this. The main thing that makes octopuses such ideal aliens is that their intelligence, which some scientists think is on par with us humans, remains foreign to us. Unlike our own, octopuses' minds aren't solely in their walnut-sized brains. Experiments have shown that octopus arms "think" independently. This occurs not just when separated from the body, but also when carrying out thoughtful actions such as grabbing food and directing movement. Each limb carries neurons which work without the command of the brain. It's hard for us humans to comprehend the idea of possessing autonomous limbs, as ours are so dependent on our central nervous system.

The contrast between human and octopus minds is an evolutionary one. Cephalopod intelligence naturally evolved under markedly different circumstances than most creatures we think of as intelligent such as apes or parrots. And nearly all the animals we consider smart are long-lived, communal creatures, those whose intellect was at least in part spurred on by the need to make complex bonds with other members of their species. Human minds evolved to recall individuals, remember who helped and who didn't, and sustain bonds that last for lifetimes.

But cephalopods live brief, solitary lives. Even large species, such as the huge Pacific octopus, live for only a few years. They have little to do with other octopuses save for the need for breeding. And so octopus minds didn't evolve to form social bonds or lifelong relationships. We have little idea of why they're so smart or what evolutionary conditions led to their relative intelligence. Their intellect, like their boneless bodies, is wholly alien, even though they are from our home world. In so many ways, it makes them the perfect "monster" for the movies.

DUNE AND PANDORA: WHEN DOES THE ECOLOGY OF AN ALIEN WORLD BECOME A CHARACTER?

"Since the publication of Rachel Carson's *Silent Spring* (1962), and the awakening of the environmental movement, science fiction writers have confronted the dilemma of the effects of human beings on our biosphere. The first planetary ecology novel on a grand scale was Frank Herbert's *Dune* (1965). The ecosystem with its food web is depicted in intriguing detail, including the complex life cycle of the [giant] sand-worm which produces the unique psychotropic 'spice.'"

—*The Cambridge Companion to Science Fiction* (2003)

"When you understand that under capitalism a forest has no value until it's cut down, you begin to understand the root of our ecological crises."

—Adam Idek Hastie, Twitter (2020)

DUNE AND PANDORA: LIFE HELL, LIFE HEAVEN

Dune: The desert planet Arrakis. For some, a hell planet. Those few life-forms that survive, like the giant sandworms, must do so on very little water. Pandora: A pastoral planet, replete with life. A heaven so profound that the ecology forms a huge neural network spanning the entire lunar surface onto which the creatures can connect. Even though some scientists still debate the nature and breadth of life, there is little doubt that, like Earth, planet Pandora is teeming with life. And the more we look, the more we explore, the more Earth seems to teem.

Organisms, known as extremophiles, thrive on the Earth's ocean floor. Such extremophiles survive under high pressure and temperature on the rocks near so-called hydrothermal vents. Extremophiles can also live deep down in ice caps. In the Antarctic ice shelf, more than 300 miles down from surface nutrients and Sunlight, these stationary animals survive. Geologists recently found a new species of extremophiles while drilling through the Antarctic ice. And extremophiles can even live in orbit, miles above our heads. An experiment onboard the International Space Station showed that the bacteria *Deinococcus radiodurans* can survive at least three years in space. Every last nook and cranny of our planet harbors life of some sort.

PANDORA: LIFE HEAVEN

That's the kind of ecology conceived for Pandora. In James Cameron's 2009 movie *Avatar*, the Pandoran biosphere is bursting with alien life. Cameron used a team of expert advisors to make Pandora's fauna and flora as realistic as can be. They created a biodiversity of bioluminescent species. Bioluminescence is the making and releasing of light by a living organism. On Earth, bioluminescence occurs commonly in marine fauna, as well as in some selected fungi and terrestrial fireflies.

On Pandora, the biodiversity ranged from hexapodal (six-legged arthropod) animals to tropical-type flora which are several times taller than that existing on Earth. Most Pandoran plant and animal species have bioluminescent properties. The flora seen in the movie were designed by Jodie Holt, a professor of botany at the University of California, Riverside.

Professor Holt's idea was that the Pandoran flora are able to communicate with each other through signal transduction, the process by which a chemical or physical signal is transmitted through a cell. And the flora are larger than on Earth because of the greater atmospheric thickness, weaker gravity, and stronger magnetism on Pandora.

With regard to signal transduction on Earth, scientists have recently become more aware of the deep networks between tree "families." Trees can send each other carbon through fungal threads. This process isn't random. It's a *deliberate* sending process. Research shows that mother trees don't just prioritize their offspring with key nutrients. They also send water, nitrogen, phosphorous, defense signals, and allele chemicals through these fungal networks. Mother trees can be connected to hundreds of trees in this way, and the goodness they pass to those trees increases seedling survival fourfold. Research is crucial to conservation. If too many mother trees are cut down, the whole system collapses. These fungal networks, with their nodes and links, make it clear that there's a *wood wide web* in action. Fungi act as links, trees as nodes. The busiest nodes are the mother trees.

PANDORAN FAUNA

Pandora is a moon, of course. In the movie, Pandora is said to be located in the Alpha Centauri A system, just over four light-years from Earth. It is one of the many natural satellites orbiting the gas giant Polyphemus, named for the Polyphemus of Greek mythology. As the moon has less gravity than Earth, the animals' larger sizes match their environment.

James Cameron's initial idea for Pandora's fictional fauna was for them to be superslick and aerodynamic. Designers then tried to put flesh on the bones of this idea with three main conceptions of the fauna design: that they should appear expressive, that they should function with animation technology, and that they should seem realistic to an audience who have, after all, a huge amount of experience observing fauna on planet Earth.

Aiming for superslick designs that drew from seashells, turtles, and insects, the designers found that the main challenge was to give the fictional creatures organic appearances, especially skin texture. Some of the animals were designed to include breathing holes located in the trachea, similar to spiracles in some of Earth's animals. Another design challenge was animal locomotion. Imagine having to design a "walk and run" cycle for a six-legged creature. Not easy. Another challenge was to portray credible flight technique for creatures that have four wings. Indeed, many of the creatures also have four eyes, with a major eye/minor eye coupling on either side of their head.

The Pandoran animals are not connected telepathically. Although the idea of the creatures being tuned into Pandora's "world-mind" was discussed, in the end they settled for some "lesser" animals as having heightened instincts. It's an interesting way to show that the human invaders on Pandora are somewhat *other* compared with the natives. Pandoran natives are hooked up to their planet. It's the humans who are detached and don't properly appreciate the huge ecological wealth of the Pandoran biosphere.

PANDORAN PARADISE

What makes some planets like Earth and Pandora a "paradise" for life? Water. The water cycle dominates our climate on Earth and is the perfect solvent for the chemistry of life. As water stays stable over a big temperature range, Earth is at just the right distance from the Sun for water to remain a liquid solvent. And, even though the first movie focused on Pandora's land, *Avatar 2* will focus on the underwater world of Pandora's oceans.

As James Cameron put it, "Part of my focus in the second film is in creating a different environment—a different setting within Pandora. And I'm going to be focusing on the ocean, which will be equally rich and diverse and crazy and imaginative, but it just won't be a rainforest. I'm not saying we won't see what we've already seen; we'll see more of that as well." This rainforest theme is an interesting one for a paradise planet. During the Carboniferous period of Earth history between about 360 and 300 million years ago, our planet went through what's known as the Age of the Coal Swamps. It's when life on Earth really started to thrive, with great tropical rainforests, giant dragonflies and amphibians, as well as the first reptiles.

DUNE: LIFE HELL

If Earth is some form of paradise, then Venus is a hell. Planetary scientists used to think Venus was Earth's sister planet. Maybe a giant swamp planet, like Earth's Carboniferous past. But when we sent word to Venus, our probes melted in minutes. Any newcomer brave enough to visit Venus would be roasted by the heat, crushed by the pressure, and poisoned by the atmosphere. Venus really wouldn't make much of a holiday destination. As weather on a planet is made by its atmosphere, little wonder Venus is so hostile. Our sister planet has an atmosphere ninety times thicker than ours, with high winds, plenty of volcanoes, and blazing heat. (There are places on Earth almost as hell-like as Venus. The Atacama Desert in Chile is one of the driest places on Earth, fifty times drier than the Sahara Desert. Atacama has no plants, no animals, not even bugs in the soil. It's so dry that even bacteria can't survive—just like Venus.)

In Frank Herbert's *Dune*, water is also lacking. Though no way as bad as Venus, life on Arrakis is harsh. As Herbert writes, "The effect of Arrakis on the mind of the newcomer usually is that of overpowering barren land. The stranger might think nothing could live or grow in the open here, that this was the true wasteland that had never been fertile and never would be." The book explores the impact people have on an ecosystem, and vice versa. Herbert carefully crafted his *Dune* world to be a case study in ecological cause and effect.

Planet Arrakis is almost completely covered in arid deserts. Here and there, the odd scattered mountain range sometimes break up the seemingly infinite seas of sand. The mountains provide shelter to the limited native life-forms that call *Dune* their home. Like Pandora, who have the Nav'i, Arrakis has its own humanoid species in the Fremen. The Fremen tribes use the rocky mountains and caves as places of residence, as deep within exist vast reserves of water.

PLANET ARRAKIS IS UNIQUE

Like Earth, the atmosphere of Arrakis is mostly made up of nitrogen and oxygen. This makes the *Dune* world naturally breathable by humans. It's the metabolism of the mighty sandworms that produce the oxygen, enabling other oxygen-breathing biological life to survive. Tiny amounts of water exist as vapor in the atmosphere. This is harvested by the Fremen using wind traps, which are essentially large air intakes, allowing the moisture to be captured and condensed and funneled into a catch basin.

The *Dune* world is unique. It is the only planet where the substance known as spice is found. In Herbert's novel, the entire infrastructure of civilization across deep space depends on spice. It's the substance that extends life and makes interstellar travel possible. The dramatic and dominant life-form on the *Dune* world is the massive sandworm. And spice is actually a by-product of the life cycle of the sandworms.

Ultimately, it is the sandworms that have transformed this *Dune* world. The huge creatures are the key to why Arrakis sits at the center of all power and influence in this fictional Universe. Not only that, but the fight for survival in this hell world was the dominant force that formed the Fremen's cultural identity. The brutal *Dune* ecology meant the Fremen had to be vigilant in their use of energy and resources, especially water. And this hugely contributed to the Fremen's emergence as doughty warriors, able to use their skills to defend their planet from tech-advanced off-worlders.

ECOLOGY IS THE STAR

Alien narratives in which ecology plays a central character, not merely in the setting, but in the very plot, are rare. Stories that do this well are even rarer. And yet *Dune* and *Avatar* succeed. For writers and moviemakers, one of the main challenges in creating a believable alien ecology is that everything in an ecosystem interacts with everything else. In nature, whether on Earth or on some alien world, there are food chains. There will always be more small creatures and plants than large dominant ones, a fact almost always overlooked by writers for the sake of drama and action. It's as if, in the great majority of alien fiction, in tales where the ecological background is ignored, the aliens exist in a vacuum.

Dune has been called the first planetary ecology novel on an epic scale. Herbert's themes of environmental cause and effect are heavily threaded throughout the story. Arrakis wasn't always a hell planet. The discovery of large salt flats by imperial paleontologists suggests that the planet once had lakes and oceans. And Herbert's work proved to have a progressive impact on our own society. Environmentalists have pointed out that *Dune*'s portrayal of a planet as a complex and living entity strongly influenced environmental movements, such as the initiation of International Earth Day.

Avatar is a tale of indigenous resistance to ecological destruction. Earth tries to colonize distant planets to satisfy its resource hunger. In this mission, humans find Pandora to be an environmentally rich planet with a diverse and profoundly elaborate ecosystem, a paradise. The indigenous Na'vi defend their home world, engaging with humans over what is, in essence, an ecological question: What's more important to life, robbing the world's resources or the balance and harmony of its nature?

In conclusion, these two tales, *Dune* and *Avatar*, use ecology as a central character so well that we as spectators draw parallels between the environments of these fictional worlds and the way in which resources are greedily and mindlessly harvested on Earth today.

IS JAR JAR BINKS TO BLAME FOR SO FEW AQUATIC ALIENS?

"Mesa called Jar Jar Binks. Mesa your humble servant . . . I spake!
. . . How wude! . . . Mesa cause one, two-y little bitty axadentes,
huh? Yud say boom de gasser, den crashin der bosses heyblibber,
den banished . . . Ay-yee-yee! Wha! Was'n dat. Hey, wait! Oh,
mooie-mooie! I love you! . . . Noah gain! Noah gain. Da being
hereabouts, cawazy! Wesa be wobbed un crunched! . . . Ohh,
maxi big, da Force. Well, dat smells stinkowiff . . . Ex-queeze-me."
    —Jar Jar Binks, *Star Wars: Episode I – The Phantom Menace*
(1999)

Herman Melville: "How does one come to know the unknowable?
What faculties must a man possess? It has pushed man to venture
further and further into the deep blue unknown. We know not
its depths, nor the host of creatures that live there. Monsters.
Are they real? Or do the stories exist only to make us respect
the sea's dark secrets?"
                                                    —*In the Heart of the Sea* (2015)

TOP TEN MOST ANNOYING MOVIE CHARACTERS

In this wired-up world of clickbait and meme thread, you're never very far away from online polls. You know the kind of thing: Top Ten Largest Body Parts; Top Ten Ways to Avoid Texas (sorry, that should have read "Taxes"); Top Ten Satanic Things to Take to a Church Picnic. And so on. Naturally, one of the most popular topics of such lists is the movies: Top Ten Movies about Extraterrestrials; Top Ten Highest-Grossing Movies (dominated by science fiction films, of course); Top Ten Movies Where the Villain Wins (Google this last list).

Down among these jabber lists, where the signal-to-noise ratio is almost always low, the persistently bored surfer can often find polls such as Top Ten Most Annoying Movie Characters. Over the years, countless characters have ruined otherwise great films. Maybe it's the script. Maybe it's the actor. Or maybe it's both. But sometimes a character just doesn't click, no matter how much the moviemakers try.

So, what annoying characters are conjured up from cinema's past from such countdowns? Jim Carrey's histrionic Riddler from *Batman Forever*? Chris Tucker's annoying and androgynous radio DJ Ruby Rhod from *The Fifth Element*? The Ewoks, surely? But one constant character in all this conjuring is the infamously idiotic Jar Jar Binks in *Star Wars: Episode I – The Phantom Menace*.

MARITIME SCIENCE FICTION

In conjunction with silly old Jar Jar, and while we're on the topic of clickbait lists, consider how very few notable works there are in the subgenre of maritime science fiction. The subgenre begins with Jules Verne's 1870 classic *Twenty Thousand Leagues Under the Seas*, but then has very little of note, save perhaps John Wyndham's *The Kraken Wakes* and Michael Crichton's *Sphere*.

This lack of maritime science fiction is surprising. Consider planet Earth. "Earth" is a curious choice of name for a home world that is around 71 percent covered in oceans. Given that another 1 percent of our planet's surface is covered by lakes and rivers, around only 28 percent of the world could be called "Earth." In many ways, we live on planet Water. The wonderful variety of life on Earth *depends* on water. Life on our planet is thought to have begun in the oceans. And water makes up 60–70 percent of all living matter. About two-thirds of the human body is composed of water, and humans can live no more than a week without a drink (not including whiskey). Water is the "matrix" of life. It's the environment in which life thrives. It's Earth's lifeblood, coursing and pumping through our planet with the ingredients for life. And when scientists go looking for life beyond the Earth, they follow the water, as it were. As NASA puts it, "The hunt for water continues to drive exploration today, even beyond Earth. Water may exist on other bodies in the Solar System, like Mars and the Moon."

EUROPA CREEK WITH NO PADDLE

Look at Europa and you'll see why water excites the scientific imagination. Jupiter's system is like a Solar System in miniature. Last time we looked, Jupiter has at least 79 Moons in orbit about it, just like the planets orbit about the Sun. Europa is one of the four important innermost Moons, discovered by Galileo in 1610.

Europa is a water world. It's slightly smaller than our Moon but, as this lunar world is around 485 million miles away from the Sun, its surface is a cracked and twisted patchwork of ice. From above, Europa's surface looks a little like the Earth's poles, with solid ice floating on water. But the big water question that scholars want answered is whether there's a huge ocean beneath this frozen crust. In future missions to Europa, space agencies may send down a probe to drill through the icy crust and see what lurks beneath.

The deep dark ocean that might lurk beneath Europa's icy crust would be very chilly indeed. Even from Earth we can spy the 125-mile-high jets of liquid that erupt from Europa's south pole, so not only might there be an ocean under that crust, but its water is being shot out directly into space. If scholars have the math right, Europa's seas have twice the water found on Earth. The mean temperature of Europa's "sea" is around -256°F. But that doesn't mean it would be all ice. The gravity of mighty Jupiter makes huge tides inside Europa. These tides make the jets we can see from Earth. But the tides also mean that Europa is being warmed up, maybe just enough to make those seas liquid. Scholars wonder if Europa might be warm enough for life to start and stay in those seas. If, like Earth, Europa has hydrothermal vents, then maybe the chemistry is right for alien sea creatures to lurk in those dark Europan waters.

TWENTY THOUSAND LEAGUES UNDER THE SEAS

The maritime sci-fi subgenre got off to a reasonable start with Jules Verne's *Twenty Thousand Leagues Under the Seas*. The title refers to the actual distance traveled under the various seas. It doesn't refer to any depth achieved in the story itself, as twenty thousand leagues (almost fifty thousand miles) is almost twice the circumference of the Earth. The greatest depth reached in the tale is a mere four leagues, about ten miles, but even that is almost three miles deeper than the ocean's maximum depth. The confusion arises due to the book's title; correctly translated, it should read *Twenty Thousand Leagues Under the Seas*, not *Sea*, with the book using metric leagues, which are four kilometers each. In Verne's story, it's the year 1866 and seagoing vessels of various nationalities spy a mysterious sea monster. The American government assembles an expedition, centered around the frigate *Abraham Lincoln*, to find and destroy the beast. Ultimately (spoiler!) they are startled to find that the "monster" is in fact a futuristic submarine, the *Nautilus*. The book's account of Captain Nemo's *Nautilus* is ahead of its time, as it correctly describes many features of today's submarines, which in the 1860s were relatively primitive affairs. And the rest of the story depicts the protagonists' adventures aboard the *Nautilus*—built in secrecy and now lurking beneath the seas, beyond the reach of land-based governments or authority.

Though *Twenty Thousand Leagues* is not directly an alien fiction, there is nonetheless sufficient evidence of just how "other" the Earth's oceans are in terms of the real possibilities for portraying real alien worlds and life-forms. The ocean's deep remains one of the most mysterious places on our planet. We know less about these locations than we do about the surface of Mars, say. More people have walked on the Moon, twelve in total, than visited the bottom of the Mariana Trench, which lies at a depth of 35,853 feet. Four thousand people have been to Everest, and almost six hundred to space; only seven have made it to Challenger Deep. Freezing cold, bible black, and with crushing pressures, the deepest part of the ocean is one of our world's most hostile realms. What kind of creatures manage to survive there?

FINDING GUNGANS, NOT NEMO

Given how hard it is to imagine the unimaginable when it comes to alien life, wouldn't an ocean setting give a writer or moviemaker a head start? As the deep is such a mysterious place, the creative imagination could run riot. But what have we to show for all this potential? The alien monster in *Twenty Thousand Leagues* turns out to be a futuristic submarine. The aliens in John Wyndham's *The Kraken Wakes* merely land in Earth's oceans, as they can only survive under conditions of extreme pressures in which humans would be instantly crushed. And Michael Crichton's *Sphere* barely counts, as it's a story about a spacecraft of unknown origin found on the bottom of the Pacific Ocean.

Then, there's the Gungans in the Star Wars Universe. The Gungans can breathe underwater. They are a semi-aquatic humanoid species indigenous to planet Naboo. They're tall with a bill-like mouth, large nostrils (that close when diving), and long fin-like ears that help with swimming. They have transparent membranes over their eyes that allow them to see underwater. Gungans are omnivorous with strong teeth and a long extendable tongue with which they capture food. That might sound mighty weird, but the Gungans are amphibian in nature. They start life as tadpole-like critters. As aquatic creatures, Gungans prefer a moist environment, as arid weather sears their skin. They're able to jump powerfully as their skeleton is cartilaginous and very flexible. Gungans have human-like dexterity, as they have four fingers and an opposable thumb on each hand. Despite their amphibian nature, Gungans evolved as land creatures, as witnessed by their cloven feet and humanoid body plan.

In the checkered history of maritime sci-fi, the conception of the Gungans is a kind of ballpark Darwinism, a hotchpotch of body parts somehow thrown together by nature. The ideas are no more advanced than that of Greek thinker Empedocles who lived almost five centuries BC. Consider evolution, Empedocles style. For no apparent reason, body parts are swimming around in Earth's primeval oceans. On initial combination, the elements produce some strange results: heads without necks, arms with no shoulders (and, we could add, ears which are also fins). In time, as these simple scraps of anatomy coalesce, horned heads on human bodies emerge. In the same way, bodies of oxen with human heads evolve, and creatures of twin sex. Generally, however, these peculiar products of natural forces perish as quickly as they come (unless they coalesce to make a Jar Jar). More rarely, bits and pieces of anatomy adapt to each other, and more complex creatures live on. So was the organic Universe formed, sprung from spontaneous aggregations. Including, we might add, the Gungans.

THE TRUE MEANING OF *SOLARIS*

All this rather wanting, maritime sci-fi makes you marvel once more at *Solaris*, Stanisław Lem's 1961 novel. The story of *Solaris* follows a crew of scientists on a research station as they attempt to understand an extraterrestrial intelligence. The ET intelligence takes the form of a vast ocean on the titular alien world. And yet, in perfect keeping with the terrestrial mysteries of oceans, the intelligent entity is designated as if it were an ocean. But is it an ocean? Perhaps it's a brain, or a protoplasmic machine—a gelatin. The terrestrial scientists visiting this alien world know it is none of these things. And the library of the Solaris station displays in a thousand volumes the shattered attempt to overcome the "impossibility . . . that the reality might be totally alien."

Terrestrial science lacks an objective or meaningful point of reference. The real meaning of "ocean," "plasma," "gelatin," or "brain" lose all meaning in the context of this alien world. And so, communication among the scientists themselves becomes a problem. The true meaning of the entity under study begins to flow away. All that was solid about terrestrial science melts into ocean, as it were, under the metallic colossus of their research station. How can the scientists expect to communicate with the ocean when they don't even understand one another?

IT'S NOT JAR JAR, IT'S US

The problem with aquatic aliens has a backstory. When Copernicus deprived the Earth of its central position in the cosmos, turning it into a mere planet in orbit about the Sun, he opened up the possibility of the material similarity between our world and alien worlds. If the Earth is a planet, then the planets may be Earths. The Universe was no longer thought of in terms of otherness. And alien worlds became graspable, if only in terms of being physical worlds like the Earth. With the discovery of our own world bursting open with voyages of discovery, the search for knowledge was directed not only to new places on Earth but also outward into new worlds in space. And yet, one location on Earth that remained relatively unexplored was the ocean. Extraterrestrial invention needs a terrestrial analog on which to base the creative imagination. And we still know so little about the seas. So, no, Jar Jar Binks is not to blame.

WOULD LUKE SKYWALKER REALLY HAVE BEEN HUMAN?

"Science can, and does, strive to grasp nature's factuality, but all science is socially embedded, and all scientists record prevailing 'certainties,' however hard they may be aiming for pure objectivity. Darwin himself, in the closing lines of *The Origin of Species*, expressed Victorian social preference more than nature's record in writing: 'As natural selection works solely by and for the good of each being, all corporeal and mental endowments will tend to progress toward perfection.' Life's pathway certainly includes many features predictable from laws of nature, but these aspects are too broad and general to provide the 'rightness' that we seek for validating evolution's particular results—roses, mushrooms, people, and so forth."
—Stephen Jay Gould, *The Evolution of Life on Earth* (1994)

MEANWHILE, ON TATOOINE

Tatooine. A harsh desert world orbiting twin Suns in the Galaxy's Outer Rim. A lawless planet ruled by gangsters, while settlers eke out a living on moisture farms, and spaceport cities serve as a jump-off point for pirates, smugglers, and actors wearing silly costumes. Humans. The Star Wars Galaxy is riddled with them. What else is new? Science fiction is chockablock with human "aliens."

Luke Skywalker is human. As is Leia Organa. And Darth Vader, natu-rally. Obi-Wan Kenobi? Human. Han Solo? Yep. Padmé Amidala, Qui-Gon Jinn, Rey, and Finn, all human. Of course, the presence of humans in the *Star Wars* story is important dramatically. Nobody disputes that fact. The very first movie in the franchise establishes their vulnerability and weakness early on. Luke Skywalker's introduction early in the first movie was rewritten to enable this.

It's the job of science-minded people to ask the awkward questions. So, let's ask one: How are there humans in the Star Wars Galaxy? They're on planets Naboo, Alderaan, and Tatooine, among others. How did they get there? According to *Star Wars* canon, the human species are native to the planet Coruscant. In the fictional Star Wars Galaxy, Coruscant sits among the populous Deep Core worlds, the Galaxy's central and most luminous region of space. This is where humans first evolved and then migrated out once they were granted permission to settle new planets. Straightforward enough? Well, there's a problem even with this simple scenario.

HUMANS ON EARTH

We've already seen how on our modest little world, life seems to have found its way into every last little nook and cranny. Life abounds in jungles, thrives in rivers and seas, and flourishes in forests. Even deserts and cities are stuffed with wildlife. Almost all habitats are jam-packed with bugs and beasts of every known detail and description. We can assume that billions of years before Coruscant was a planet-wide metropolis and capital of the Galactic Republic, it too was thriving with life.

But life on Earth hasn't always been like this. Most scientists believe that life came ashore from the oceans at least half a billion years ago. To provide you, dear reader, with something of a potted history of evolution (given that most of you are limbed animals), most scholars still believe that life began in the oceans (there is an increasingly vocal minority who believe life began not in the sea, but on land). Roughly speaking, tetrapod (four-footed) creatures like us evolved from lobe-finned fishes, which began to come ashore those many millions of years ago. And that's how we believe most large land animals evolved. Of course, some, like the blue whale, returned to the sea. It is believed that Coruscant was once an Earth-like world, mostly covered in oceans, so we would have to assume that the most unlikely coincidence happened there too: Coruscant tetrapods came ashore.

This leap from the sea to the land shows that once life has taken hold, it can make big jumps in its development, yet this rather simple view obscures the fact that we believe unicellular life reigned supreme for more than two billion years on Earth before the evolution and spread of multicellularity. These unicellular critters were micro-beasts that lived in all parts of the world wherever there was water: in soil, hot springs, around hot vents on the ocean floor, high in the atmosphere, and deep inside rocks within the Earth's crust. Of course, we can't say the same for fictional Coruscant, but the presence of humans would require life, a similar kind of flora and fauna within a breathable atmosphere.

THE AGE OF THE MICRO-BEASTS

On Earth, as we have seen, life can survive the most extreme places. Our planet's two icy polar oceans are very inhospitable places, though tens of thousands of species have been found there. But first prize goes to a microscopic, eight-legged organism known as the water bear, or tardigrade. It survives the cold, −459°F, and the heat, 303°F. It endures one thousand times more radiation than other animals and can live without water for ten years. It is the only known creature to survive the airless vacuum of space.

Micro-beasts ruled our planet for billions of years. Did they do so on Coruscant too? On Earth, it was only about half a billion years ago when more complicated creatures arose during a period known as the Cambrian explosion. This "explosion" was the rather rapid appearance, over many millions of years, of most major groups of "complex" animals. Did Coruscant have a similar evolutionary explosion? No one can doubt that more complex creatures arose after Earth's long unicellular beginning. First, eukaryotic cells, perhaps about two billion years ago, then multicellular animals about six hundred million years ago, with a spread of peak complexity among animals passing from invertebrates to marine vertebrates, and finally to reptiles, mammals, and humans. Life would have to have gone along a similar pathway on Coruscant, even though the potential for evolutionary difference between the two worlds would have been huge.

THE GUT OF LUKE SKYWALKER

Now, the convention for representing evolution in the old charts and texts as a progression that runs "age of invertebrates," followed by "age of fishes," "age of reptiles," "age of mammals," and "age of Man" (adding the antiquated gender bias to all the other bigotries wrapped up in this sequence) is very outdated. Humans have an understandable need to view history as progressive and to view ourselves as predictably dominant; what's sometimes called "the becoming of Man." You can see that a similar assumption is made in the creation of the fictional world of Coruscant.

Scientifically speaking, this bias has hugely corrupted the true view of life's pathway on this planet. It foolishly places as its focus a relatively minor phenomenon that arises only as a side consequence of a physically constrained starting point. In short, we big up humanity way too much. In fact, and without hyperbole, the primary feature of terrestrial life has been the stability of its bacterial mode. From the very start of the fossil record until you read this sentence and, no doubt, into all future time so long as the Earth abides, this is actually the "age of bacteria." So it was, and so it ever shall be. Witness the fact that there are more Escherichia coli cells living in your gut right now than the number of humans living on this planet. The same would be true of Luke Skywalker's gut.

DON'T GO LOOKING FOR HUMANS IN SPACE

So, maybe we should look for micro-beasts when exploring other Galaxies, far, far away, and not humans. Though there may be millions or billions of Earth-like planets out in space, their speed of life might be slower, so the life pathway on other worlds may still be in the "micro-beast" stage of its development. Organisms adapt to and are constrained by physical principles. It is, for example, scarcely surprising, given laws of gravity, that the largest vertebrates in the sea (whales) exceed the heaviest animals on land (elephants today, dinosaurs in the past), which, in turn, are far bulkier than the largest vertebrate that ever flew (extinct pterosaurs of the Mesozoic era).

It's highly unlikely that humans on another planet like Coruscant evolved to out-compete other native life forms, eventually overrunning and urbanizing the whole planet. And the likelihood of them being genetically similar to humans is impossible, given the huge number of random events that took place on Earth to allow humans to exist in our current form. For Coruscant humans to be the same species as us, we would have to occupy the same branch on the tree of life.

At the risk of truly annoying *Star Wars* fans, *Star Trek* did a much better job with the whole "humans in space" thing. In an episode of *Star Trek: The Next Generation* called "The Chase," Captain Picard discovers that there is an embedded genetic pattern in all humanoids, which is constant throughout many different sentient species, and which was left by an early ancestor race that predates all other known civilizations in the Star Trek Galaxy. Now *that* might work.

WHAT DO ALIENS IN DC'S *SUPERMAN* AND MARVEL'S *AVENGERS* HAVE IN COMMON?

"And we men, the creatures who inhabit this Earth, must be to them at least as alien and lowly as are the monkeys and lemurs to us. The intellectual side of Man already admits that life is an incessant struggle for existence, and it would seem that this too is the belief of the minds upon Mars. Their world is far gone in its cooling and this world is still crowded with life, but crowded only with what they regard as inferior animals. To carry warfare Sunward is, indeed, their only escape from the destruction that, generation after generation, creeps upon them."

—H. G. Wells, *The War of the Worlds* (1898)

MARTIANS AND EARTHLINGS

Earthling Mark Watney invades Mars in the 2015 movie *The Martian*. Watney finds that, to make a colony on the Red Planet, humans would have to fend off global dust storms, big swings in temperature, harmful solar radiation, and maybe melt the Martian polar ice into a sea that is covering much of the planet.

One hundred and seventeen years earlier, the Martians had invaded Earth. H. G. Wells's novel *The War of the Worlds* was sci-fi's first ever "menace from space," first ever alien invasion. In Wells's tale, Mars is a dying world. Its seas are evaporating, its atmosphere dispersing. The entire planet is cooling, so "to carry warfare Sunward is, indeed, their only escape from the destruction that generation after generation creeps upon them."

So, the terror of the void is brought to Earth. In his story, Wells gives his readers repeated reminders of "the immensity of vacancy in which the dust of the material Universe swims" and invokes the "unfathomable darkness" of space. Life is portrayed as precious and frail in a cosmos that is largely deserted. Wells's book also carefully conveys the quality of the void—immenseness, coldness, and indifference—in its rendering of the aliens. His Martian machines vividly hammer home the cosmic chain of command: "It is remarkable that the long leverages of their machines are in most cases actuated by a sort of sham musculature . . . Such quasi-muscles abounded in the crablike handling-machine . . . It seemed infinitely more alive than the actual Martians lying beyond it in the Sunset light, panting, stirring ineffectual tentacles, and moving feebly after their vast journey across space."

As we have seen, the modern alien owes almost everything to Wells. His Martians are agents of the void. Their strange physiology and intellect make them the prototypical alien. Their Tripods tower over men physically, as the vast intellects of their occupants tower over human intelligence. Bodily frail, but mentally intense, the Martians and their superior machines are instruments of human persecution. Their weapons of heat rays and poison gas are dehumanizing devices of mass murder. All attempts at contact are futile, furthering the idea of the aliens as an unrelenting force of the void.

ALIEN INVASIONS

Ever since Wells, it seems, aliens have been forever invading our humble little planet. One of the more recent from many famous examples is the Marvel Cinematic Universe's battle of New York, which features in the 2012 movie *The Avengers*. Also known terrestrially as "the Incident," the battle of New York was a week-long clash of titans, with the Earth and the

Avengers on one side, and Loki and the Chitauri army on the other. But that's just the latest example. Marvel comics also gave us the subversive, long-term secret invasion of Earth by the shape-shifting alien Skrulls in 2008.

Alien invasion has become an inexhaustible topic for fantastic film and fiction, but the sophistication of alien depiction has developed little since *The War of the Worlds*. Wells's novel says nothing about Martian culture. It seems to have wasted away by some entropic decay and whittled down to nothing more than a cosmic justification to invade Earth. The Martians have no interest in human culture. Like vampires, they are interested only in human blood. And this clinical rationale for human oppression inspires readers' loathing of the Martians, and the latent power of unsolicited natural selection.

Whereas Wells's Martian invasion seems scientifically justified, the vast majority of other alien invasions are not. At least Wells's Martians are inhabitants of a dying planet, which is fast winding down into desert. But this exceptional case for alien invasion, within our Solar System, has been thoughtlessly adopted for a whole genre. Writers of film, fiction, and comic books have mostly aped the Wellsian invasion myth, but without the exceptional case Wells made for the Martians. Not only that, but in an attempt to eclipse the master and his Martians, writers attributed ever greater power to the aliens. And they gifted to aliens the promise of unimaginable riches, a glittering prize of planet Earth not only of value to the small desert world of Mars, but for any imaginable civilization in the Galaxy and beyond.

Consider the 2011 movie *Cowboys & Aliens*. And, yes, it's quite apparent that a student of alien science should be wary from the get-go, based on the film's title alone. The aliens in the movie are not only abducting people to conduct experiments on them (this is standard fare by now and gets no less ridiculous), but also mining gold. As American film critic Roger Ebert put it at the time: "The aliens, as usual, show limited signs of intelligence. Oh, they arrive in a spaceship that's taller than a skyscraper, and they must have designed it. But mostly they strafe the town, drop explosive charges behind characters, but rarely upon them, and reel up human victims into their smaller flying ships in order (need we be told) to study them. Their other purpose in journeying unimaginable distances across the void is to use mysterious forces to suck up gold—coins, watches, rings, whatever." Of course, one can't complain about aliens wanting gold. Gold is very nice. Why they needed gold isn't really explicitly explained in the film, but how did the aliens know the gold was here, on our planet, probably billions of miles from where they live?

KAL-EL COMES CALLING

Another alien invasion is that of Superman. The Man of Steel was famously born as Kal-El on the alien planet of Krypton. His parents, Jor-El and Lara, learn of Krypton's imminent destruction, sounding exactly like the kind of entropic decay H. G. Wells envisioned for Mars. So, Jor-El starts to build a spacecraft to carry Kal-El to, guess where? Why, Earth, of course. Now, according to *Superman #132*, planet Krypton was three million light-years away from our world. Cosmically speaking, that's not so far, but it's hardly next-door neighbor either. Over *twenty other Galaxies* sit between Superman's world and the Milky Way.

What, one might wonder, inspired Jor-El to send his son Earthward? Let alone the exacting logistics of launching a rocket on a light-years trajectory in any direction. Given that there are certainly billions of planets between Krypton and here, and assuming there are probably (Drake willing) thousands of civilizations per Galaxy, you get a good idea of the lethargic thinking behind the plan.

Like *The War of the Worlds*, the Superman story has no culture clash between Earthlings and aliens. Clark Kent is raised in secret. There is no great revelation that he's an alien from a sophisticated and advanced civilization, which is perhaps just as well, as it would make the reader wonder why Jor-El's race of supermen had even bothered with our humble little world. Astronomers recently estimated that there are a dizzying two trillion Galaxies in the Universe. That's up to twenty times more than was previously thought. And this recent revelation was based on 3D modeling images collected over two decades by the Hubble Space Telescope. Even if there are only a thousand civilizations per Galaxy, that puts the number of advanced civilizations in the cosmos at two thousand trillion. With all this real estate in the cosmos, Earth would have to be very exceptional to justify a visit. Don't you think?

Things look even more incredulous when you consider the invasions of what we might call superhero aliens. Powerful armies, with vast and monstrous masses of starships at their disposal, are all hell-bent on taking over our tiny world. Technology has sharpened their cosmic fangs, all the better to eat humanity. And yet the difference in scale is like assuming an Earthly superpower, such as China, is dead set on mobilizing its armies to expropriate the local corner store. In reality, the readers know that the cost of invasion must be worth more than the value of the booty.

THE SUPER-ALIEN COSMOS

Yet you never really know with the super-alien cosmos. Maybe the aliens simply aren't motivated by material gain. Perhaps these super-aliens attack our planet merely because it pleases them to do so. They destroy for the sake of destruction. They enslave humanity as an amused and academic exercise in despotic mastery. Shame, really. It's a far cry from Wells. *The War of the Worlds* was an exercise in interplanetary Darwinism. Wells's imaginative lens was like a type of telescope with the invading Martians being the "men" of the future. But it was the wrong end of the telescope— Imperial Britain was on the receiving end of social Darwinism.

Wells's wrath was focused on the becoming of Man and the prevailing idea at the time that polite English, middle-class society was the very point of evolution. The target of Wells's parody was the same kind of flabby thinking that Mark Twain superbly satirized in his wonderful essay *Was the World Made for Man?*: "Man has been here 32,000 years. That it took a hundred million years to prepare the world for him is proof that that is what it was done for. I suppose it is. I dunno. If the Eiffel tower were now representing the world's age, the skin of paint on the pinnacle-knob at its summit would represent man's share of that age; and anybody would perceive that that skin was what the tower was built for. I reckon they would. I dunno."

SEEKING EARTHLINGS TO SWAT

The vast majority of alien invasion stories since Wells show little evidence of planning, imagination, or genius. Wellsian interplanetary Darwinism seems to have been replaced by a kind of paranoid philosophy penned by the Marquis de Sade. This imagined paranoid cosmos seems obsessed with the conquest of (mostly harmless) Earth. It is a cosmos that sets every type of trap to catch humankind, whether by outright attack or by stealth and robbing us of our free will (a scenario that proved very popular during the days of the Cold War and the Senator Joseph McCarthy years).

The paranoids and hyper-vigilants are with us still. They apply the same logic to the SETI or METI debate about whether or not it's okay to send messages to nearby stars in the hope of attracting the attention of their hypothetical inhabitants. In short, whether to beam "yoo-hoo" messages to ET. No doubt having read the repeated and monotonous alien invasion scenarios, borrowed from Wells but now showing a poverty of philosophy, the paranoids fear aliens will invade because (a) they sadistically feel like it, (b) they fancy a cosmic game of cops and robbers, or (c) they are a kind of intergalactic band of Jehovah's Witnesses or Scientologists coming to save us from ourselves.

When it comes to invading super-aliens, it's high time for a change. Humans are no angels and would surely have no qualms about killing a cockroach. But they would hardly go to the ends of the Earth to do so. Likewise, aliens, if you'll forgive the comparison of humans to cockroaches. Such super and sophisticated civilizations would have no need to go out of their way, seeking Earthlings to swat. And *that's* science.

WHY MIGHT OUR GALAXY BE FULL OF EWOKS?

"Forget about love. Scientists want to know: Are we looking for life in the wrong places? An astrophysicist at the UK's University of Lincoln says that exoMoons, Moons which orbit planets outside of our Solar System, could contain liquid water and therefore, support life. 'These Moons can be internally heated by the gravitational pull of the planet they orbit, which can lead to them having liquid water well outside the normal narrow habitable zone for planets that we are currently trying to find Earth-like planets in,' Dr. Phil Sutton said in a statement. 'I believe that if we can find them, Moons offer a more promising avenue to finding extra-terrestrial life.'"

—Fox News, "ExoMoons could be home to extraterrestrial life, researcher says" (2019)

ALIEN SPLITTERS

There are few aliens who split opinion more than Ewoks. Their very name inspires a range of knee-jerk emotions in science fiction aficionados. In the positive blue corner, we have those that view the bear-like creatures, who first materialized on the silver screen in 1983's *Return of the Jedi*, as a charming part of sci-fi's rich tapestry of tropes. In the negative red corner, there are those who consider the Ewoks a ridiculous bunch of Baloos, a pack of Paddingtons, or a wad of Winnies and would prefer they not exist at all. The Ewoks are splitters, the most controversial inhabitants of the Star Wars Universe, bar good old Jar Jar Binks.

And that's because the Ewoks live on Endor, the franchise's most famous exomoon. This so-called Sanctuary Moon has been among the most dramatically important planetary settings, as the battle of Endor saw the revolution to restore the Republic. It was on this Forest Moon that the Imperial forces kept their shield generator, which at first prevented the Alliance from destroying the Death Star.

According to canon, Endor was around three thousand miles across—easily bigger than our Moon which is a little over two thousand miles in diameter. Endor was roughly 43,000 light-years from its Galaxy center, orbited a gas giant planet called, er, Endor, and covered in dense woodlands. The Moon had a breathable atmosphere, and along with its planetary companion orbited the two Suns . . . Endor I and Endor II (the franchise working overtime on names, there). But what does Endor and its Ewoks mean for alien life?

The idea of life on planetary Moons is not new. For hundreds of years people believed there was life on our Moon. Some thought the lunar craters were actually forts, and others thought they saw winged bat people through their telescopes. But relatively recent exploration of the Solar System by robots has revealed the Moons to be real worlds in their own right. And that's got us all thinking there may be fish swimming in icy alien seas of Europa, or tiny critters clambering over the organic muck on Saturn's moon Titan. After all, gas giant planets such as Jupiter, Saturn, and Endor are mini planetary systems. They each have Moons in orbit about them. And these Moons follow the same facts of life as the planets. If the conditions are right, life might evolve. Even if it is freezing out there.

THE GOLDILOCKS ZONE AND THE RED DWARF

In planetary systems such as Endor, scientists think that a planet must lie in the "habitable zone" to support life. This is well reported by now, and by "habitable zone" they mean the range of various orbital distances around a star within which rocky planets can support liquid water on their surfaces. If a planet is beyond the outer limits of the habitable zone, it will not get enough of the Sun's energy and water will freeze. If a planet is within the inner limits of the habitable zone, it will get too much stellar energy and surface water will boil away. This not-too-hot, not-too-cold idea is why scientists sometimes call the habitable zone the Goldilocks Zone. In our Milky Way Galaxy and fictional Galaxies such as that in *Star Wars*, looking for alien life isn't just the simple matter of looking for exoplanets. Planet hunters must also narrow their searches to aim for Goldilocks Zones. And, in that search, red dwarf stars become important because of exoMoons like Endor. Here's why.

Our Galaxy contains between one hundred and four hundred billion stars, although the actual number may be as high as one trillion. Around the orbits of these stars exist at least one hundred billion planets, many of them Earth-like. Now, according to estimates, red dwarfs make up 75–80 percent of the stars in the Milky Way. They are by far the most common type of star. The same can also be assumed of the Star Wars Galaxy, but that proves an interesting problem for life on planets in orbit about red dwarfs.

RED DWARFS

A red dwarf is a small, cool star. Such stars have masses less than half of the Sun's. Red dwarf stars were once thought unable to support habitable planets, but scientists now think that old theory is wrong. Red dwarfs *do* have planets, but they have curious orbits. Planets in the Goldilocks Zone of a red dwarf would be so close to the parent star that they would be tidally locked. Just like the Moon in orbit about the Earth, the red dwarf's planets always show the same face to their Sun.

So red dwarf worlds are locked by the parent star's gravity. One half of the planet is always in darkness, the other half always bathed in light. They are strange and exotic worlds. On the dark side, there is a vast frozen waste. On the light side, oceans, a temperate climate, and land. However, there is also a "Twilight Zone," an in-between place. Here, between the contrasting half-worlds, strange creatures may compete for food and light. If there are any creatures at all, in such an alien world.

But even if a red dwarf exoplanet doesn't harbor life, its exomoon may. Imagine an Earth-sized moon in orbit around a Jupiter-sized giant. The giant may be locked in its orbit around the red dwarf, suffering extremes of climate, yet the moon in orbit around the giant may be habitable. It would circumvent the tidal lock problem by becoming tidally locked to its planet. This way there would be a day/night cycle as the moon orbited its primary, and there would be distribution of heat.

MANY ENDORS?

This is why exomoons like Endor may be so important. Since the rather humble red dwarfs predominate in Galaxies, the planets and Moons in orbit about them predominate as well. As red dwarfs make up three-quarters of the stars in the Milky Way, their planets and Moons may even prove to be the norm. And, as the planets themselves may be inhospitable half-worlds, their Moons may be a common home of life in the Galaxy.

Also, consider the lifetimes of red dwarfs. They are huge. A red dwarf half the mass of the Sun has a lifetime of fifty-six billion years. At the moment, it's not clear whether the energy output of red dwarfs is stable enough for the development of life. But if civilizations do develop on red dwarf exomoons, they could have huge lifetimes, too. It's possible that the meek truly will inherit the Galaxy.

Could Endor be the greatest of all *Star Wars* insights? Could Ewoks be a common form of alien life? Endor is a habitable forest moon in orbit around an inhospitable gas giant. The diameter of Endor is about 40 percent of the Earth's. How common are such worlds in our Milky Way? Until we explore further, we can only wonder at the chances that civilization lives on an exomoon.

PART III
ALIEN WORLDS

ALIENS ON OUR MOON: WHAT WAS AMERICA'S GREAT MOON HOAX?

"By means of a telescope, of vast dimensions and an entirely new principle, the younger Herschel . . . has already made the most extraordinary discoveries in every planet of our solar system; has obtained a distinct view of objects in the Moon . . . has affirmatively settled the question whether this satellite be inhabited, and by what order of beings; has firmly established a new theory of cometary phenomena; and has solved or corrected nearly every leading problem of mathematical astronomy."

—*The New York Sun* (1835)

NEWSPAPERS AND *THE NEW YORK SUN*

By the dawn of the 1800s the scientific vision of progress had begun to be realized. And one of the most rapidly expanding areas of industry was newspaper publication. In the early 1800s there were over fifty newspapers in London alone, and another one hundred titles in the rest of the UK. As tax duties were reduced from the 1830s onward, there was a massive explosion in circulation, as the newspapers fed public hunger for news. In the post-revolutionary United States, the story was the same. Publications spread like wildfire. In 1800 there were up to two hundred titles, by 1810 over 350, and during the following two decades the increase was equally rapid.

In this climate emerged perhaps the most remarkable alien newspaper report ever published: the discovery of a civilization on the Moon. This "hoax" began in August 25, 1835, when the *New York Sun* carried the first of six installments professed to be a lunar revelation by the eminent astronomer, Sir John Herschel. So breathtaking was the detail of the initial report that over nineteen thousand copies of the August 26 issue of the *New York Sun* were sold. It was the largest circulated newspaper on the planet.

The publishers smelled success and resounding sales. The August 29 issue of the *Sun* declared that the entire report was to be sold as a pamphlet. Within days, sixty thousand copies of the pamphlet had been sold. And the booklet has made a number of re-appearances since. Illustrations of the lunar aliens were also sold. And various translations of this extraordinary publication popped up around the globe. Even by the close of 1836 there had been Italian translations in Florence, Naples, Ravenna, and Livorno; French translations in Paris, Strasbourg, Lyon, Bordeaux, and Lausanne; a German translation in Hamburg; and Spanish translations in Mexico and Cuba.

THE GREAT MOON HOAX

This much of the "hoax" was true. In 1835, English astronomer John Herschel *was* at the Cape of Good Hope, as the newspaper reports suggested. But the publication from which the report was allegedly culled, namely the *Edinburgh Journal of Science*, was no longer in print. The *New York Sun* kept its readers waiting. For the most part, the first installment was a mere description of Herschel's telescope, which boasted a twenty-four-foot mirror capable of forty-two thousand times magnification. But with the second issue on August 26, the "observed" lunar world was unveiled. First, a tantalizing glimpse of the Moon's geology and botany, then Herschel's scribe, Dr. Andrew Grant, declared "our magnifiers blest our panting hopes with specimens of conscious existence." They spy a being like a bison, which is seen to have " one widely distinctive feature, which we afterwards found common to nearly every lunar quadruped we have discovered; namely, a remarkable fleshy appendage over the eyes, crossing the whole breadth of the forehead and united to the ears. We could most distinctly perceive this hairy veil, which was shaped like the upper front outline of the cap known to the ladies as Mary Queen of Scot's cap, lifted and lowered by means of the ears. It immediately occurred to the acute mind of Dr. Herschel that this was a providential contrivance to protect the eyes of the animal from the great extremes of light and darkness to which all the inhabitants of our side of the Moon are periodically subjected."

Their delight knows no bounds as they spot a bearded hircine beast, blessed with a prominent horn. No longer have they spied a flock of stunning birds and a shoal of lunar fish than the installment ends with the setting of the Moon. The next issue on August 27 kept the readers waiting with reports of lunar volcanoes and seas, but the jackpot came the following day when Herschel and his co-observers reported seeing creatures who "averaged four feet in height, were covered, except on the face, with short and glossy copper-colored hair, and had wings composed of a thin membrane, without hair, lying snugly upon their backs, from the top of the shoulders to the calves of the legs. The face, which was of a yellowish flesh-color, was a slight improvement upon that of the large oran-utan, being more open and intelligent in its expression, and having a much greater expansion of forehead . . . In general symmetry of body and limbs they were infinitely superior to the oran-utan; so much so, that, but for their long wings, Lieut. Drummond said they would look as well on a parade ground as some of the old cockney militia!"

The question of alien lunar intelligence was swiftly solved when the creatures are observed "evidently engaged in conversation; their gesticulation, more particularly the varied action of their hands and arms, appeared impassioned and emphatic. We hence inferred that they were rational beings, and, although not perhaps of so high an order as others which we discovered the next month on the shores of the Bay of Rainbows, that they were capable of producing works of art and contrivance."

BAT-MEN

The creatures are then christened "man-bat," and gifted with the accolade of intelligence: "We scientifically denominated them the Vespertilio-homo, or man-bat; and they are doubtless innocent and happy creatures, notwithstanding some of their amusements would but ill comport with our terrestrial notions of decorum." When the August 29 issue was published, *Sun* readers were treated to reports of lunar oceans and a "magnificent . . . temple—a fane of devotion, or of science, which when consecrated to the Creator, is devotion of the loftiest order." The series of reports was brought to an end with the last installment on August 31, climaxing at a high point—a description of yet higher creatures and a telling of the "universal state of amity among all classes of lunar creatures." A brief appearance is made by Saturn, but Herschel's calculations are rather patronizingly left out of the report "as being too mathematical for popular comprehension." As the series is brought to an end, the highest species of bat-men are spied: "In stature, they did not excel those last described, but they were of infinitely greater personal beauty, and appeared, in our eyes, scarcely less lovely than the general representation of angels by the more imaginative school of painters."

US ABUZZ

The United States was abuzz with this "discovery." On September 1, the *New York Sun* reported on the reactions of other publications to these lunar alien revelations. The *Mercantile Advertiser* declared that "It appears to carry intrinsic evidence of being an authentic document." The *Daily Advertiser* triumphantly suggested "No article, we believe, has appeared for years, that will command so general a perusal and publication. Sir John has added a stock of knowledge to the present age that will immortalize his name, and place it high on the page of science." *The New York Times*, according to the *Sun*'s account, decreed the discoveries "probable and plausible," while *The New Yorker* decided they ushered in "a new era in astronomy and science generally."

Predictably, the church was not entirely amused. One contemporary account says "Some of the grave religious journals made the great discovery a subject of pointed homilies." Another report tells of an American clergyman who forewarned his congregation that he would start up a fighting fund for bibles to donate to the lunar, and no doubt heathen, bat-people. Even the question of slavery cropped up. The Slavery Abolition Act passed through the UK Parliament in 1833 and had abolished slavery throughout most of the British Empire. During the Great Moon "Hoax" it was claimed that "the philanthropists of England had frequent and crowded meetings at Exeter Hall, and appointed committees to inquire . . . in regard to the condition of the people of the Moon, for purposes of relieving their wants . . . and, above all, abolishing slavery if it should be found to exist among the lunar inhabitants."

EDGAR ALLAN POE POPS UP

There were eyewitness reports of the "hoax." It seems some academics were not sufficiently skeptical of the *Sun*'s coverage: "Yale was alive with staunch supporters. The literati—students and professors, doctors in divinity and law—and all the rest of the reading community, looked daily for the arrival of the New York mail with unexampled avidity and implicit faith." Eminent American author Edgar Allan Poe was later to report "Not one person in ten discredited it, and (strangest point of all!) the doubters were chiefly those who doubted without being able to say why—the ignorant, those uninformed in astronomy, people who would not believe because the thing was so novel, so entirely 'out of the usual way.' A grave professor of mathematics in a Virginia college told me seriously that he had no doubt of the truth of the whole affair."

It was neither the first nor last alien controversy, of course. There was a public controversy following the publication of Galileo's observations through the telescope in the early 1600s. Here again was a controversy emanating from novel, if not Earth-shattering, observations. And the same drama would be played out again with the Martian canals controversy in the late nineteenth century, and the UFO farce of the mid-twentieth.

THE BUBBLE BURSTS

At last, though, the bubble burst on the Great Moon "Hoax." A journalist from the *Journal of Commerce (JoC)* was dispatched to the *New York Sun*. His mission: to secure a copy of the full Moon report so that the *JoC* may consider republication. The reporter met with a *Sun* staff writer by the name of Locke who told him, "Don't print it right away. I wrote it myself." The *JoC* subsequently accused the *Sun* of publishing a hoax, and the *New York Herald* named Locke as the architect.

And so Richard Adams Locke finally enters the frame. Locke was an American descendant of the Newtonian philosopher John Locke. Some accounts have Richard Adams Locke studying at the University of Cambridge, though Cambridge has no evidence he ever did so. He did, though, work as a writer and editor in England before setting up in New York, where eventually he joined the *Sun* in 1835. According to several biographies, Locke seems to have been interested in science, which is resoundingly confirmed by the impressive detail about astronomy in his *Sun* articles. Edgar Allan Poe had Locke down as a skillful writer. Just prior to 1835, Poe had written his own lunar fiction, so he was well placed to comment on Locke's talent in writing with "true imagination."

DOCTOR DICK

Critics have made a compelling case that the Great Moon "Hoax" was in fact merely satire. They present "solid historical evidence" that Locke used his science skills at the "delicate task of writing satire." Locke's target was all believers in alien life, but Thomas Dick in particular. Dick had been a Scottish churchman, but also a science teacher and writer. He was well known for his works on astronomy, as well as his attempts to defuse tension between the two fields.

Not everyone was pleased by Thomas Dick. In his 1852 account of the "Hoax," William Griggs was critical of Dick's work: "It would be difficult to name a writer who, with sincere piety, much information, and the best of intentions, has done greater injury, at once, to the cause of rational religion and inductive science, by the fanatical, fanciful, and illegitimate manner in which he has attempted to force each into the service of the other, instead of leaving both to the natural freedom and harmony of their respective spheres." Griggs had no doubt that Dick was the subject of Locke's infamous spoof: "We have the assurance of the author, in a letter published some years since, in the New World, that it was written expressly to satirize the unwarranted and extravagant anticipations upon this subject, that had been first excited by a prurient coterie of German astronomers, and thence aggravated almost to the point of lunacy itself, both in this country and in England, by the religio-scientific rhapsodies of Dr. Dick. At that time the astronomical works of this author enjoyed a degree of popularity, in both countries, almost unexampled in the history of scientific literature. The scale of the editions republished in this country was unbounded until nearly the whole of his successive volumes found a place in every private and public library in the land."

TALKING TO THE BAT-MEN

Griggs pressed the point even further. He confirmed that the articles had been penned as a result of Locke's reading material during the summer of 1835. He had been reading an 1826 issue of the *Edinburgh New Philosophical Journal*. And in this issue Thomas Dick had reported the idea of constructing vast geometrical signals, by which Earthlings could communicate with "selenites," as the assumed inhabitants of our Moon were known. The strikingly unusual nature of this early idea of interspecies communication bears a closer look.

The architect of the selenite signals was Carl Friedrich Gauss. Gauss was something of a colossus in nineteenth-century science. Sometimes known as the "Prince of Mathematicians," and even the "greatest mathematician since antiquity," he was also onetime director of the Göttingen Observatory in Germany. Gauss appears to have been a believer in lunar life, a belief that he was allegedly keen to pursue in a scientific culture that was increasingly skeptical about evidence of life on the lunar surface.

Gauss's plan was this: send signals to the Moon or Mars, so that extraterrestrials would know we were clever. His preferred methods of sending such signals were ambitious and dramatic. Method one: in the Siberian forests, set up a vast "windmill" diagram as used in Euclid's illustration of Pythagoras's theorem. Naturally, the method assumed, the selenites would respond in kind, as math on the Moon was just the same as that on Earth. The lunar men would realize that Earth was inhabited and draw fitting conclusions.

Method two was no less dramatic: the construction of a huge canal, cut into the Saharan desert. Once the canal had been cut the climax of the master plan would be realized: onto the canal waters paraffin would be poured, and the whole thing set ablaze. Using this method, a different signal could be sent every night. Once more the selenites would be stirred by such a striking display of terrestrial "intelligence." This second method seems to have been the brainchild of a colleague of Gauss's, one Johann Joseph von Littrow, the director of the Vienna Observatory.

SIGNALS CROSSED

Alas, as with most things associated with Richard Adams Locke, the selenite signal story appears to be apocryphal. Not only is there many a different telling of the tale, but there is also no narrative actually containing reference to the writings in which Gauss and Littrow are originally meant to have set down their respective plans of lunar action. And yet the account acts as a perfect introduction to the wild and wonderful world of Thomas Dick. Indeed, the sheer potential farce of the ideas led Locke to regard the science as "fair subjects of sedate and elaborate satire," as Griggs was to quote.

But the signals story was just a beginning. Locke then opened the works of Thomas Dick himself and found it hard to believe what he was reading. One thing was for sure, however: Thomas Dick's back catalog was ripe for lampooning. Griggs was not alone in considering Locke's articles satirical. At the French Académie des Sciences, astronomer François Arago read them aloud to other academy members among "repeated interruptions from uncontrollable and uproarious laughter." John Herschel had mixed feelings. At first, he laughed along with those in the know. But later, after receiving questions from believing English, German, French, and Italian enquirers, the farce was not so funny.

HOAX OR SATIRE?

There appears to have been some academic debate on whether Locke's "Hoax" was indeed a satire. But this was finally laid to rest in 2004 when James Secord, a prominent science historian at Cambridge, located the original letter, dated May 6, 1840, that Locke sent to Griggs explicitly saying the articles were satirical. Locke wrote that he "had become convinced that the imaginative school of philosophy . . . was emasculating the minds of our studious youth," and weakening them for normal science. He added: "One of the most conspicuous of the jingling heads of this school, is the famous Dr. Dick of Dundee, who pastes together so many books about the Moon and stars, and devoutly helps out the music of the spheres with the nasal twang of the conventicle. It was this ciphering sage's 'Christian Philosopher' that suggested the Moon-story."

Ironically, Locke's satire failed. And yet that very fact is telling. Locke had miscalculated the mood of a gullible generation raised on the works of Thomas Dick and his disciples. As revolutionary industrialist William Greggs put it in 1852, "The soil had been thoroughly ploughed, harrowed, and manured in the mental fields of our wiser people, and the seed of farmer Locke bore fruit a hundredfold."

Locke's satire illuminates the problems for scientists in the early nineteenth century, as they were caught on the horns of a dilemma: religion on one side, and lack of scientific evidence on the other. Locke's spoof uncovered the naïveté of the general readership. It exposed the overstated claims of its authors and struck a much-needed note of caution in the debate. Nonetheless, the Great Moon "Hoax" remains perhaps the most entertaining episode in the search for alien life.

SPACEBALLS: HOW DO ALIENS AVOID LIFE'S BIG OBSTACLES?

"The chunks of comet Shoemaker-Levy 9 were so large, and were moving so fast, that each hit Jupiter with at least the equivalent energy of the dinosaur-killing collision between Earth and an asteroid 65 million years ago. Whatever damage Jupiter sustained, one thing is for sure: it's got no dinosaurs left."

—Neil deGrasse Tyson, *The Sky Is Not the Limit* (2000)

"Needles in a heavenly haystack. There are more stars in the heavens than human beings on Earth. Through telescopes, men of science constantly search the infinitesimal corners of our Solar System seeking new discoveries, hoping to better understand the laws of the Universe. Observatories dedicated to the study of astronomy often are set in high and remote places. But there is none more remote than Mount Kenna observatory in this part of South Africa. 'If our calculations prove to be correct this will be the most frightening discovery of all time. These two bodies have traveled almost a million miles in two weeks.'"

—Sydney Boehm, screenplay, *When Worlds Collide* (1951)

ARMAGEDDINGOUTAHERE!

You know how it is. One moment you think your species is doing just fine, eking out a niche in your planet's environment, then—*KABOOM*—disaster strikes. Literally. And the worst kind of disaster is one that comes from outer space. A big obstacle like an asteroid or comet just smacks into your world and ruins life as you know it.

Cosmic impacts can wipe out life. Throughout our history, back to the beginning of planet Earth itself, impacts have been a problem. Impacts have been so common that the *deep time* of Earth's past is organized into a time scale that uses these big impacts to mark the beginning and end of each time period. Change on Earth isn't all slow and gradual. Life on Earth has evolved in creeps and revolutionary leaps. And there have been some very dramatic leaps. The young Earth collided with another proto planet to form the Moon, and around sixty-five million years ago a comet or asteroid hit Earth and wiped out lots of life.

Since life began on Earth, there have been a number of particularly huge so-called extinction events. These occur when conditions on Earth have led to a sharp fall in the variety and abundance of life. The most spectacular of these was the time of the "Great Dying," around 252 million years ago. Then, 96 percent of all marine species and 70 percent of all land species became extinct. The big extinction event of sixty-five million years ago was, of course, the one that killed the dinosaurs, and was also an event in which 75 percent of species became extinct. Planetary scientists believe that such big impacts reduced Sunlight to the Earth and made life difficult, leading to a catastrophe in the Earth's ecology.

THE END OF THE ALIEN WORLDS

Science fiction is full of ways of ending the Earth. But no one seems to think about how those same dangers that face our planet would also be faced by all other planets in the Universe. At least if we are right, of course, that planetary systems are pretty much the same when you compare one part of the cosmos to another. In short, what's good for Earth should also be good for the home worlds of bug-eyed monsters, big and small.

So let's look at six ways of killing ET even before they have the chance of invading poor old Earth.

1: SOME DREADED EXTRATERRESTRIAL LURGEE

Sci-fi has an obsession with plagues, an obsession which goes all the way back to Mary Shelley. As Shelley sat writing *Frankenstein* during 1816, a book that some say is the first work of science fiction in the modern age, the skies outside were darkened by volcanic ash during that so-called "year without a summer," following the eruption of Mount Tambora in Indonesia. Shelley's second work of science fiction directly follows the apocalyptic events that might have been set in motion by the Tambora explosion. *The Last Man* (1826) deals with the effects of a worldwide plague, which drifts across the face of the globe leaving just a sole survivor.

The horror of Shelley's vision of an annihilating plague was brought to life with the 1918–19 Spanish Flu pandemic (and gave us all more than a few worries during the most recent pandemic). An estimated 5 percent of the world's population died from the Spanish Flu, around 100 million souls, making it one of the deadliest natural disasters in human history. Science fiction needed no further encouragement. The last century was dominated by movies such as *The Andromeda Strain* (1971), *28 Days Later* (2002) and its sequel, *28 Weeks Later* (2007). The portrayal of future pandemic apocalypses has influenced the World Health Organization to plan and coordinate responses to future outbreaks across the world. And now we have more than enough warning.

But have you ever thought about an extraterrestrial plague? If the conditions of life are the same on alien worlds, then alien races would also be vulnerable to disease. Remember that the all-conquering Martians in *The War of the Worlds* were finally killed not by humans but by an onslaught of Earthly pathogens, to which they had no immunity. As H. G. Wells wrote, "slain, after all man's devices had failed, by the humblest things that God, in his wisdom, has put upon this Earth."

2: WARFARE

H. G. Wells was hugely influenced by Mary Shelley's work. His own vision of the explosive devastation caused by an atomic bomb became the defining focus of concern about the end of the world for most of the twentieth century. Visions of atomic apocalypse were inspired by the infamous detonations in Japan in 1945, and the ensuing Cold War and its test explosions. Sci-fi movies such as the adaptation of Nevil Shute's relentlessly morbid *On the Beach* (1959) capture the post-war mood of impending doom. The story is set in Australia, the only continent left unscathed by a nuclear war. Here too the fallout would soon engulf the land, and as the action ends with a poignant closing shot, a banner flutters in the breeze to a soundtrack of "Waltzing Matilda." It reads: THERE IS STILL TIME . . . BROTHER.

Alien civilizations too would be subject to warfare. Indeed, there's war aplenty in Robert A. Heinlein's *Starship Troopers*, made into a movie in 1997, and sci-fi television series such as *Battlestar Galactica* and *Star Trek*, in which, for example, the Dominion War is a conflict between the forces of the Dominion, the Cardassian Union, and, eventually, the Breen Confederacy against the Alpha Quadrant alliance of the United Federation of Planets, the Klingon Empire and, later, the Romulan Star Empire.

3: ALIEN WORLD CRISIS

As well as surviving planetary war, aliens might also have to deal with environmental crises. On Earth, atomic power is the plot of thrillers such as *The China Syndrome* (1979), *Silkwood* (1983), and the television miniseries *Chernobyl* (2019), which explore cover-ups at nuclear power stations and speak of the risks that future societies face when they go totally nuclear. Of course, *Chernobyl* was based on the factual 1986 Chernobyl Disaster which led to a plume of radioactive debris drifting across the Soviet Union and Europe.

Tales about the demise of the world as we know it can be natural rather than technological. Notable here is John Christopher's 1956 post-apocalyptic novel, *The Death of Grass*, in which a new virus strain infects crops, causing a massive famine. The book was retitled *No Blade of Grass* for the United States, as supposedly the US publisher thought the original title sounded like something out of a gardening catalog. These days, if we mess up this planet at least we know there are other worlds out there we can try traveling to. Unless the aliens come here first. And, once again, the same could be said for any extraterrestrial civilization, if they are to survive.

4: ALIEN INVASION

Exactly how aliens know we're here on planet Earth is anyone's guess, as they must surely have come from somewhere deep in space. But maybe an alien world out there is in a more densely packed neighborhood, and the interplanetary telltale signs of life are easy to spot. The alien invasion subgenre of science fiction focuses instead on what the aliens want to do to us. But an alien civilization may also be invaded, naturally, and would have to survive the invasion or face extinction. Earthly tales of greedy extraterrestrials invading Earth either to slaughter and supersede us or to enslave us or to harvest us as food applies equally, whether silly or not, to the alien worlds themselves. (After all, as Douglas Adams jokes in *The Hitchhiker's Guide to the Galaxy*, your planet may be destroyed to make way for a hyperspace bypass.)

5: COSMIC CATASTROPHE

In his 1897 lesser-known short story, *The Star*, H. G. Wells has a strange luminous object erupt into the Solar System and make its way toward a fragile Earth. As the object looms ever closer, havoc is wreaked upon the Earth as gravitational tidal waves cause global devastation. Similar tales were told in movies throughout the next century. *When Worlds Collide* (1951) is a film about the coming destruction of the Earth by a rogue star, and so bears a close resemblance to Wells's story. In "Five Years," a song written by English musician David Bowie, released on his 1972 album *The Rise and Fall of Ziggy Stardust and the Spiders from Mars*, an impending apocalyptic disaster will destroy Earth in five years and the being who will save it is a bisexual alien rock star named Ziggy Stardust!

Sometimes, actual cosmic catastrophes inspire film and fiction. In July 1994, humans witnessed the first ever extraterrestrial collision when Comet Shoemaker-Levy 9 plowed into planet Jupiter. The collision was no doubt the inspiration for the "paranoia of the two 1998 movies, *Deep Impact* and *Armageddon*, where Bruce Willis again saves the world in his now familiar vest. Hopefully, alien worlds and civilizations also have an extraterrestrial version of Bruce Willis in a vest, as planetary systems will form and evolve in the same way the cosmos over.

6: ARTIFICIAL INTELLIGENCE

In *AI: Artificial Intelligence*, Steven Spielberg's 2001 movie, it's rising sea levels and global warming that drastically reduce Earth's population. And two millennia later, humans have become extinct, and Manhattan is buried under ice. Humans have been superseded by an advanced silicon-based intelligence known as Mecha. In *The Matrix* (1999), the demise of humans comes a lot earlier in the twenty-first century, as intelligent machines wage that traditional Frankenstein-like science fiction war against their creators. This cult movie became so popular that whenever someone was spotted on a pay phone, people assumed they were looking for an exit from the Matrix.

Once more, when considering alien worlds and their intelligent inhabitants, it could be that the beings themselves have died out but their intelligent tech remains. Artificial intelligence, rather than natural intelligence. In Carl Sagan and Iosef Shklovskii's 1966 book, *Intelligent Life in the Universe*, the authors consider intelligent alien civilizations. They discuss the possibility that the lifetime of a civilization is not indefinitely long. And they stress that the death of civilization on one small planet does not imply the end of intelligent life in the Universe. Sagan and Shklovskii point out that, just as one individual can introduce a concrete, if small, contribution to society, a given planetary culture may make a contribution to the evolution of intelligent life in the Universe. And, just as the contribution of an individual in society would be nothing without its communication, the cultural contribution of a planet to the evolution of intelligent cosmic life simply cannot happen without interstellar communication. The aliens would live on through their tech.

HOW DOES ALIEN MIGRATION HAPPEN IN *STAR WARS* AND *STAR TREK?*

"Humanity's interest in the heavens has been universal and enduring. Humans are driven to explore the unknown, discover new worlds, push the boundaries of our scientific and technical limits, and then push further. The intangible desire to explore and challenge the boundaries of what we know and where we have been has provided benefits to our society for centuries. Human space exploration helps to address fundamental questions about our place in the Universe and the history of our Solar System. Curiosity and exploration are vital to the human spirit and accepting the challenge of going deeper into space."

—NASA, *Why We Explore* (2021)

GALACTIC LIFE ZONES

If an alien civilization can survive life's big obstacles and avoid extinction on its home world, it is likely to think next about expansion and colonizing local worlds. As American science fiction author Larry Niven once said, the dinosaurs became extinct because they didn't have a space program. Larry was being quoted by Arthur C. Clarke, the author of *2001: A Space Odyssey*. Clarke had been in conversation with Buzz Aldrin, the second man to walk on the Moon. The two futurists were contemplating the real human space odyssey in the century that lies ahead. Niven's point was this: if the dinosaurs had had a base on the Moon or Mars, they'd have had a better chance of survival.

So how would alien migration occur? Before we take examples from *Star Wars* and *Star Trek,* let's consider some Galaxy science. Now, an alien world doesn't just need to be in the Goldilocks Zone 'round its mother Sun to support life. Astrophysicists and planetary scientists believe that a planet also has to be in the right part of a Galaxy, a system of stars in which most Suns live. Just like planetary systems in orbit about stars, Galaxies too have their own life zones where the development of life is more favorable. So, a planetary system must be in the right neighborhood of a Galaxy. Scientists think that if the alien world was too close to the galactic center, evolving life would suffer far too much from the harmful radiation in that zone. But it can't be too far out in a Galaxy either.

In the case of our Milky Way Galaxy, its habitable zone contains about six billion stars. It is commonly believed to be a chubby ring with an outer radius of about 33,000 light-years and an inner radius *close* to the Galactic Center (with both radii lacking hard boundaries). "Alien" Galaxies will have larger or smaller life zones, depending on their characteristics and makeup. Most Galaxies are one of three types: spirals, ellipticals, or irregulars. This taxonomy is based on the way a Galaxy looks, what's known as Galaxy morphology. Other important features are a Galaxy's rate of star formation, and how active its core, or center, is. The latest estimate of the number of Galaxies in the Universe is two trillion, so there is a lot of real estate in deep space.

In space science, astrophysicists define a heavy element as anything heavier than the simplest elements of hydrogen and helium. Heavy elements are important materials in the formation of rocky planets like Earth, and scientists think they are needed to make complex life. Some parts of Galaxies are richer in heavy elements than others, so from this point of view, too, the idea of a galactic habitable zone makes sense.

Finally, Galaxies rotate. Astronomers have discovered that all Galaxies rotate at least once every billion years or so, no matter how big they are. The Earth spinning around on its axis once gives us the length of a day, and a complete orbit of the Earth around the Sun gives us a year. Our Sun, along with Earth and the rest of the Solar System, orbits the center of the Milky Way Galaxy. And it takes the Sun around 250 million years to orbit the center of the Milky Way once, an orbit which is known as a cosmic year. And regardless of whether a Galaxy is very big or very small, if you could sit on the *extreme* edge of its disk as it spins, it would take you about a billion years to go all the way around.

ALIEN MIGRATION IN *STAR TREK*

How does alien migration occur in the *Star Trek* and *Star Wars* Galaxies? In both cases we are dealing with humanoid aliens, though that fact says more about the lack of creative imagination about what aliens might look like than anything else. With regard to *Star Trek*, alien migration is mostly explained in the *Star Trek: The Next Generation* episode "The Chase."

Captain Jean-Luc Picard finds that there is a genetic pattern embedded in all humanoid aliens, and which is constant throughout many different sentient species. The genetic pattern is the work of an early ancestor race (whom we shall call "The Founders") that predates all other known civilizations in the *Star Trek* Galaxy. This pattern explains why so many sentient races in the Galaxy are humanoid. Along with others, Picard finds a recording of a female Founder. She explains that it is her race who are responsible for the presence of intelligent life in some parts of the Galaxy. When the Founders first explored the Galaxy, there had been no other intelligent life-forms, so they seeded numerous planets with their DNA to create a legacy of their existence after their race had passed. And *that* means the ensuing migration vectors formed a pattern which was wholly predetermined by a single sentience, the Founders. The female Founder ends her message by declaring her hope that the knowledge of a common origin for all humanoids will help produce peace in the Galaxy.

This plot in "*The Chase*" is an example of the scientific theory of panspermia. The theory has it that seeds or spores can spread life through the Galaxy, or indeed the wider Universe. This theory also holds that very small organisms, or the biological precursors of life, can be found in the Galaxy and that they created life on Earth and other planets. What appears in "*The Chase*" is a kind of directed panspermia, a *guided evolution*.

How does this fictional migration relate to the geography of a Galaxy? Poorly, in all honesty. Little attention is paid to the galactic habitable zone. Maybe the writers relied on the idea that the Founders knew where best to pop the panspermia seeds. *Star Trek* fans often joke that maps, much like warp speeds and cosmic distances, work at the speed of plot in every *Star Trek* series. And it's not just *Star Trek*. Star maps in television shows and films are often meaningless. The *Star Trek* idea of galactic zones of influence only makes sense to humans who are used to the idea of nation states with hard borders. In reality, in a galactic setting with some kind of hyper-fast travel, a border would mean next to nothing. But at least the panspermia theory of migration in *Star Trek* explains the presence of all those humanoid aliens. That's more of an explanation than you get about another Galaxy which existed a long, long time ago, in a Galaxy far, far away.

ALIEN MIGRATION IN *STAR WARS*

The *Star Wars* Galaxy is also well populated with humanoid aliens. Coruscant is said to be the home world of these humanoids. Indeed, Coruscant is a Core World. These Core Worlds, also known as the Coruscant Core, were situated in the area of the Galaxy that bordered the Deep Core and contained some of the wealthiest and most prestigious planets. Coruscant itself, as the home world from which migration flowed, was galactic capital, both during the time of the famous Galactic Republic and the subsequent Galactic Empire.

This actually fits well with what we believe to be true about our own Galactic habitable zone—that chubby ring with an outer radius of about 33,000 light-years and an inner radius *close* to the Galactic Center. The *Star Wars* Core Worlds *border* the Deep Core and run out into the galactic arms from there. Like our own Galaxy, the *Star Wars* Galaxy is around 100,000 light-years across, about 13 billion years old, and with most of the stars of the luminous Galaxy situated in a disk of spiral arms that rotate around the Deep Core. (Meanwhile, the *Star Trek* Galaxy *is* our Galaxy, of course.)

Starting from their Coruscant home near the galactic Core, the *Star Wars* humanoid aliens then migrated out to colonize the numerous star systems upon which they now abide. Assuming they gave it sufficient thought in the first place, the *Star Wars* writers appear to have used Earth as a corollary for migration here. There are two main competing theories about the migration of humans on our planet.

The first, formerly known as the "Out of Africa" hypothesis, holds that modern humans evolved in Africa, then spread to colonize the other continents. These modern humans then supplanted the earlier hominins, such as *Homo erectus* and *Homo neanderthalensis*, that were already populating those parts of the world. *Star Wars* seems to be suggesting a kind of "Out of Coruscant" model, possibly with humanoid aliens supplanting earlier creatures on the alien worlds they colonized.

The other main theory of human migration on Earth is known as the "multiregional" hypothesis. In this alternate model, modern humans evolved locally, in different parts of the globe, from the earlier hominins that were already occupying those regions. You can easily see why the multiregional model wasn't used for the *Star Wars* Galaxy. For one thing, it would be an evolutionary impossibility for humans to arise independently on so many alien worlds, irrespective of ecology. Or, if they used a mashed-up menu with a large dollop of the multiregional hypothesis, along with a dash of panspermia, the alien humanoid seed could have been propagated to various star systems, then evolved to become the modern humans that we see living on different alien worlds. Though, finally, we would surely expect to see more variation in the humanoid alien species than we do.

HOW WOULD THE ENGINEERS IN *PROMETHEUS* KNOW WHICH STAR SYSTEMS TO COLONIZE?

"One way to make sense of the gravitational interaction between a planet and a star is to imagine a game of tug-of-war. On one side, you have the star—a massive object with a really powerful gravitational field. On the other side, you have the planet, much smaller, with a whole lot less gravity. We know who wins this game—the star. That's why planets orbit stars and not the other way around. But even though the planet is small, it still has some gravitational force. It still has an effect on its host star, even if that effect is much less pronounced than the one the star has on the planet. But two can play at the gravity game—the planet's gravity causes the star to 'wobble' around a little bit."

—NASA, *5 Ways to Find a Planet: Watching for Wobble* (2021)

THE OUTWARD URGE

In Ridley Scott's 2012 movie *Prometheus*, a species of humanoid aliens known as the Engineers touch down on, what most assume is, prehistoric Earth for a sacred ritual devoted to the seeding of intelligent life. This scene is problematic for a number of reasons. Chief among them is how the evolutionary path from the DNA strand discharged by the lone Engineer could possibly result, billions of years later, in an Engineer-looking intelligent life form, in this case humans. Pure movie "magic."

Before an alien civilization can even think about colonial expansion out to local worlds, it would save a lot of time, effort, and expense if they knew in advance what star systems to visit and which of the millions of local stars had attendant planets in orbit about them. How would the Engineer scientists tell one star system from another? What could they do in advance to scan their cosmic neighborhood for colonization?

THE ENGINEERS' HOME WORLD

The Engineers are an ancient alien race of unknown origin. Little is said of the Engineers home planet. But, since we know they seeded intelligent life on Earth, let's assume the Engineers come from the same Galaxy as us, the Milky Way. Let's also remember that the Galactic habitable zone is a chubby ring with an outer radius of about 33,000 light-years and an inner radius close to the Galactic Center.

Given that the Engineers original planet is unknown, imagine Engineer planet-hunters stargazing on an ancient home world that orbits a star bordering on the Milky Way's core. Now, on Earth, there are only around five thousand stars visible to the naked eye. As the Engineers' world sits closer to the core, let's assume that, for them, ten thousand stars are visible to *their* naked eye. Yet, beyond their sight, millions of stars still populate the deep sky.

As stars are far more luminous, the Engineers can spy no other planets and Moons of the Galaxy other than those in their home solar system. And that's due to the simple fact that exoplanets are imperceptibly faint points of light compared to their parent Suns. In terms of number, exoplanets are typically a million times fainter than their progenitor stars. ExoMoons, naturally, would be even fainter than exoplanets, so the Engineers have no chance of seeing them. Looking for them would be like looking for a match lit, at night, right next to a car headlight (assuming, that is, that Engineers have matches and cars).

ENGINEERS LOOK FOR WOBBLE

The Engineer scientists still would have known of the existence of many of the exoplanets and exoMoons in the Galaxy. How so? Good old wobbly stars again. Consider a mighty star wheeling its way through the Galaxy. Though the star is huge, especially compared to our Sun, it's not as strong and stable as it seems. A star with an orbiting planet, or planets, will move in its own wobbly orbit because of the planet(s) gravity. So, when Engineer planet-hunters see a wobbly star, they know there may well be a planet in tow. And by measuring the size of the wobble, the hunters can estimate the mass of the planet.

Imagine that the Engineer planet-hunters take a look in the direction of our Solar System. They would spy a wobbly Sun. Like all stars with planets, our Sun wobbles. The combined gravity of the planets is what causes that wobble. If it were just the Earth in orbit around the Sun, the effect would be almost negligible. But the huge bulk of Jupiter is able to yank our star a distance greater than the Sun's radius. Just due to Jupiter alone, the Sun hits a speed of around thirty feet per second and takes over ten years to repeat that cycle. That's quite a feat for a planet, even if it is a giant like Jupiter.

Only in very rare cases would the Engineers actually get to *see* the stars wobble back and forth under the gravitational influence of their exoplanets. And yet they can see the light from those stars, and moving objects shift their light. It's known as the Doppler Effect, and it's the same principle as a siren shifting its pitch up and down as the ambulance races past (assuming Engineers have sirens and ambulances). The light can shift redder or bluer depending on its motion. A light source moving toward the Engineer will look ever-so-slightly bluer, and a receding light looks a tad redder.

Even though the Engineers can't *see* stars wobble, they can still detect the nuance in its light pattern as the planet(s) cause(s) it to swing back and forth. The Engineers would know that this effect works best when the planet is directly along their line of sight. That gives the most accurate readings of planet mass. But this effect can also yield a detectable signal when it's not perfectly aligned. As long as the star has a reasonable amount of back-and-forth in their direction, the light will shift.

ENGINEERS AND THE MASS EFFECTS

In the early days of planet hunting, a wobbly star search would have been the best way for the ancient Engineers to find exoplanets (if the history of planet hunting on Earth is anything to go by). Humans have found thousands in our local solar neighborhood of the Milky Way. Terrestrial planet-hunters search for twitchy stars out to about 160 light-years from Earth. Given that mighty Jupiter is bigger than all the other Solar System planets put together, it's Jupiter that mostly makes the Sun wobble. And this "Jupiter effect" means that the Engineers may also at first find very large planets in orbit about local wobbling stars, simply because bigger planets equal bigger wobbles.

If the Engineers go looking for smaller Earth-like planets, they'll probably need another method. When an exoplanet passes in front of a parent star, we call such an event a "transit." From Earth, for example, we spy occasional transits of Venus or Mercury. The planets are seen as small black dots creeping across the Sun, as Venus or Mercury effectively block and eclipse Sunlight as they move between the Sun and the Earth.

The Engineers could detect exoplanets by searching for minute dips in starlight when a planet transits a star. Once found, the planet's orbital size can be calculated from how long it takes the planet to orbit once about the star. The sheer size of the planet can be found from how much the brightness of the star drops, and the planet's mass can be calculated by math. Then, from the orbital size and measured temperature of the star, the planet's typical temperature can be worked out. Finally, from all this data, the Engineers can answer the question as to whether or not the planet is habitable. And yet, as you may know from the Alien franchise, that's just the *start* of the story!

FORBIDDEN PLANET: HOW WEIRD ARE ALIEN WORLDS?

"Skaro, the first ever alien world visited in Doctor Who . . . is where we first meet the Daleks. They share the planet with the humanoid Thals, a race of fair-haired warriors who are in constant conflict with the more Nazi-minded Daleks. Over time, Skaro was portrayed as a steamy post-apocalyptic world, densely populated by dilapidated buildings—one of which is a huge skyscraper-sized Dalek structure. Among Skaro's few continents lies an 'island of gushing gold,' where 'jets of molten gold shoot into the air.' . . . And there's the sound of the Dalek City itself, a kind of metallic sound that gives you the impression of endless corridors on an alien planet . . . It had a strange metallic resonance, overlaid with random sounds, which could have been the metal structure flexing as parts of it expanded and contracted, or the sound of some distant activity or machinery distorting into a weird sinusoidal waveform as it traveled down the metal corridors. Viewers were left with the impression of a living, and very alien, metallic entity."

—Mark Brake, *The Science of Doctor Who* (2021)

THE GREATEST FICTIONAL PLANETS

Science fiction and film has done its very best to imagine the unimaginable when it comes to alien worlds. Consider the Doctor's home planet Gallifrey in *Doctor Who*. Red grass, silver-leafed trees, and a domed capital city that would make the Wizard of Oz envious. Then there's Klendathu in *Starship Troopers*. Like Arrakis and Tatooine, Klendathu is a desolate desert world, but unlike the others Klendathu just happens to be blue in color. Then there's Superman's home world of Krypton. In orbit about a red Sun, Kal-El grew up a normal kid, but thanks to his landing on Earth, our planet's yellow Sun made him something super. The movie *Interstellar* features, among other worlds, the icy tundra-world of planet Mann. And *The Hitchhiker's Guide to the Galaxy* has an entire planet devoted to making *other* planets. Magrathea was such a planet, until it got so wealthy from making other worlds that the rest of the cosmos became relatively poor. To save the rest of the cosmos, the inhabitants went to sleep and Magrathea disappeared.

REAL ALIEN WORLDS

What weird and wonderful *real* alien worlds do scientists believe are out in space? Great ongoing missions are being planned to find and map exoplanets, and the focus is on rocky planets like the Earth. As Arthur C. Clarke said, it has slowly dawned on us humans that there's enough real estate in the sky, enough exoplanets going around other stars, to give every person on Earth, back to the first ape-man, their very own private, world-sized home. What are these alien worlds like? Heavens? Hells? Forget, for the time being, questions about how many of those potential homes are inhabited, and by what kind of creature. Let's focus on the planets themselves.

A reminder first that, in our Solar System, there are two kinds of planets: rocky, and gassy. The basic difference is this: on rocky planets, there's somewhere to land your spaceship—the ground. But gassy planets don't have "grounds"; they just have gas. So there's nowhere to land. Rocky planets form close to the Sun, in the hot zone. Gassy planets form farther out, beyond the frost line, and way out into the cold zone.

EXOPLANET ZOO

The discovery of exoplanets is a most exciting and relatively new science. Dreaming up new planets has been a favorite pastime of science fiction writers and moviemakers, but now the Universe has them beat. The cosmos has conjured up the possibility of planets that would be hard to imagine as real. You could say scientists have found a veritable zoo of exoplanets.

CARBON-HEAVY WORLDS

One cosmic possibility in this zoo of exoplanets is the "diamond" planet. These carbon-rich worlds may be found in orbit about pulsars, stars that are the remnants of a stellar explosion. Despite the small size of these carbon-heavy worlds, they have more mass than Jupiter. And their high density leads scientists to believe that the explosion of the parent star left a planetary body of crystalline carbon, similar to diamond.

Actual diamond planets may have been detected. For example, *55 Cancri e* is an exoplanet in orbit about the Sun-like host star *55 Cancri A*. In October 2012, it was announced that *55 Cancri e* could be a diamond planet. Huge parts of the planet's mass would be carbon, much of which may be in the form of diamond because of the temperatures and pressures in the planet's interior.

Interestingly, an episode of *Doctor Who* featured a "diamond" planet. The Tenth Doctor, episode "Midnight," features an airless planet situated in the fictional system of Xion. The planet is made mostly of diamond glaciers and mountains which, unsurprisingly, humans have colonized. Naturally, many exoplanets have their downsides. And here we have a hair-raising "Midnight Creature" that lurks in the shadows of the planet where even the rays of Sunlight are deadly.

WORLDS OF FLAME

The idea of a planet plagued with flame is surely hellish. But such a world of flame is quite easily imagined. A rocky planet orbiting too close to its Sun would not just be boiling hot. The star's gravity would torment the little world, resulting in tidal forces which would bend and bow the planet's surface. Such a tormented world would be violently volcanic and covered in oceans of bubbly lava.

WORLDS OF WATER

An ocean planet, or water world, is a planet whose surface is completely covered with an ocean of water. When icy planets in the outer Solar System are formed, they're roughly half water, half rock. If icy exoplanets in other systems migrate inward, they could move to a warmer orbit and become an ocean world fit for life. Their oceans would be hundreds of miles deep with huge waves running across the surface of their seas.

GAS WORLDS

Gas worlds, like Jupiter, are thought to be common. They're often massive planets with a thick atmosphere of hydrogen and helium, as with Saturn and Jupiter, or a thin outer envelope of gas, as with ice giants such as Uranus and Neptune. Many have a dense molten core of rock.

TOPSY-TURVY EARTH

Planet hunters also believe that some rocky planets could be tipped over onto their sides. These worlds would roll like marbles around their Suns, spinning on their sides, after a giant impact had made them somewhat topsy-turvy. Such worlds would have curious climates and seasons, with the equator getting more Sunlight than the poles.

RED DWARF PLANETS

Red dwarf planets may be the most abundant planets, as there are billions of red dwarf stars in the Universe. As these stars are cool, their Goldilocks Zones are close, so their planets would be locked in orbit, always showing the same face to its Sun. One half of the planet would be in eternal daylight, the other in eternal night.

CANNONBALL WORLDS

Finally, imagine a planet almost completely made of iron. Impossible? Earth has a core of molten iron covered in a crust of rocks. Some planets could have iron cores but a crust that's been almost completely blasted off by an impact. It would leave a strange world, with oceans and atmosphere, but a very bizarre chemistry.

CLOUD CITY: IS THERE LIFE ON GAS GIANTS?

"Human judges can show mercy. But against the laws of nature, there is no appeal."

—Arthur C. Clarke, *A Meeting with Medusa* (1971)

"But what exceeds all wonders, I have discovered four new planets and observed their proper and particular motions, different among themselves and from the motions of all the other stars; and these new planets move about another very large star [Jupiter] like Venus and Mercury, and perchance the other known planets, move about the Sun. As soon as this tract, which I shall send to all the philosophers and mathematicians as an announcement, is finished, I shall send a copy to the Most Serene Grand Duke, together with an excellent spyglass, so that he can verify all these truths."

—Galileo Galilei, *Sidereus Nuncius* (1610)

NEWS FROM SPACE!

On March 15, 1610, the rather unlikely named Wackher von Wackenfels drove up in his coach to the house of the internationally famous scholar Johannes Kepler. Herr von Wackenfels was in a state of great agitation. Amateur philosopher and poet, and privy counsellor to his Holy Imperial

Majesty, Herr von Wackenfels excitedly told the mathematician of news that had just arrived at Court: a philosopher named Galileo in Padua had turned a Dutch spyglass at the heavens and discovered four new worlds. A few days later, Kepler received further astounding evidence in the shape of Galileo's brief but magnificent flyer, *Sidereus Nuncius (The Starry Messenger)*. The book signaled the offensive on the feudal order of the old Universe with a new instrument of science: the telescope.

Galileo wielded the newly invented device like a weapon of discovery. A new Universe was unveiled. Earth-like mountains and craters on an imperfect Moon. Impure spots on the Sun. Countless stars that could only be seen with the aid of a spyglass. And four Moons in orbit around Jupiter, a focus of gravitation other than the Earth. No wonder Galileo inspired Kepler to argue that the satellites of Jupiter must be inhabited. And yet, four hundred years later, we still don't know if there is life on those Galilean Moons. But what about gas giants like Jupiter? Would we find life there too?

CLOUD CITY

Suspended high among the cream-colored clouds of Bespin sits the fictional floating metropolis of Cloud City. Looking much like Jupiter, Bespin first appeared in film in the 1980 movie *Star Wars: Episode V - The Empire Strikes Back*. Cloud City was not just a mining colony that extracted gas from the depths of the giant planet. It was also a place of political sanctuary. Bespin is a gas giant about 73,322 miles in diameter (Jupiter's mean diameter is around 86,881 miles).

In speculative film and fiction, floating cities are a common theme. The floating Hallelujah Mountains in *Avatar* have many antecedents. Seaborne floating islands can be found way back in Homer's *Odyssey*, written near the end of the eighth century BC, where Homer described the island of Aeolia. Such islands also reappeared in Pliny the Elder's *Natural History* of the first century AD. Since then, fiction has evolved the idea from cities and islands that float on water to ones that float in the atmosphere of a planet, whether by tech or by some magical means.

For example, in Jonathan Swift's 1726 satire, *Gulliver's Travels*, a "flying island" appears. Made mainly of metal, and measuring four and a half miles in diameter, the floating island is magnetic. Buried inside the island is a bi-polar magnet measuring six yards long. It is held in abeyance by an intense magnetic field below the Earth's surface. As well as having a layered geology like a planet, Swift's flying island is science fiction's first spaceship. Between *Gulliver's Travels* and *Star Wars*, science fiction has speculated about many floating cities of the future, including the idea that, 10,000 years from now, a city the size of New York will be able to float several miles above the Earth's surface, where the air is cleaner and free from disease-carrying bugs.

EXOTIC WORLDS

Air is one of those things in life that we take for granted. We breathe it in, day by day, second by second, without really thinking about it. Air is the name we give to the gases in the Earth's atmosphere, which are used in breathing and photosynthesis. Earth's gravity holds on to the gases, forming a surrounding layer known as the atmosphere. The atmosphere also protects life on Earth from harmful radiation and steadies the planet's temperature.

As you might imagine, the quality of the air on alien Moons and planets is quite different. Earth's atmosphere is influenced by the very life that it supports and is made up mostly of nitrogen and oxygen. Venus and Mars, however, have atmospheres mostly made of carbon dioxide. In the beginning, planets had similar atmospheres, made from the material that formed the Solar System. But soon, each planet's atmosphere changed due to the different properties of that planet. Take Earth and Venus, sister planets. Earth has evolved to support life, but Venus appears to be a living hell. The early days on Venus were probably similar to that on Earth. Around four billion years ago, Venus may have had liquid water on its surface. But the rising levels of gases like carbon dioxide in Venus's atmosphere led to a greenhouse effect, and the water was lost.

PARTIES WITH NO ATMOSPHERE

The atmospheres of planets such as Bespin and Jupiter include a layer of gases held by gravity. If the gravity is high and the atmosphere's temperature low, the gases will be held for a long time, as on those gas giants. If not, the planets have very thin atmospheres, or none at all. Many exoplanets are at least as big as Jupiter. For many years, scientists believed that gas giants were unlikely to harbor life. They felt that their Moons were far more likely to support life. It's easy to imagine an Earth-sized moon in orbit around a massive giant. If this moon is a rocky world of oceans and continents and richer in oxygen, maybe 30 percent compared to Earth's 21 percent, then the moon air has more carbon dioxide, so plant life may be far more abundant than on Earth.

Conventionally, the idea of life on gas giants was generally disregarded. But is the concept really so far-fetched? Is it really so unlikely that life could evolve in the atmosphere of a world such as Jupiter or Bespin? Maybe an alien plant that doesn't need soil but can still photosynthesize could act as the base for a food chain on a gas giant. It's also plausible that more complex organisms could survive on electrical energy from the lightning storms that often occur on worlds like Jupiter.

As water is a matrix for life, at least on Earth, maybe organisms on gas giants get water from vapor in the air. Perhaps they are able to somehow leech water out of the atmosphere, hydrating themselves just by breathing. It's even possible that some gas giants are made up of breathable air. Or perhaps the life-forms in gas giants are anaerobic (non-breathing), just like some bacteria on Earth, and have never been dependent on oxygen.

ALIENS ON GIANTS

This idea of life on gas giants has respected scientists as advocates. Both American astronomer Carl Sagan and British physicist Stephen Hawking have suggested that life could exist in the atmospheres of giants like Jupiter and Bespin. But what form would such life take? What would these alien creatures actually look like? Some scientists believe that gas giant life would possibly take one of two forms. The first form is that of flying creatures, maybe similar to that on Earth. The second option is in the form of drifting balloon-like organisms that likely use photosynthesis. This is mere speculation, naturally, as alien life on gas giants may turn out to be so much stranger.

As we speculate on what life might be like for such creatures, consider the wonderful words of American science fiction author Ursula K.

Le Guin about jellyfish in the ocean: "Current-borne, wave-flung, tugged hugely by the whole might of ocean, the jellyfish drifts in the tidal abyss. The light shines through it, and the dark enters it. Borne, flung, tugged from anywhere to anywhere, for in the deep sea there is no compass but nearer and farther, higher and lower, the jellyfish hangs and sways; pulses move slight and quick within it, as the vast diurnal pulses beat in the moon-driven sea. Hanging, swaying, pulsing, the most vulnerable and insubstantial creature, it has for its defense the violence and power of the whole ocean, to which it has entrusted its being, its going, and its will." Of course, the analogy with Le Guin's words is not exact, but the pressure in the atmospheres of gas giants would be important to any living creature, as altering its altitude would give it a distinct advantage. And the most energy efficient way to do this would be through buoyancy. That means either keeping an internal density similar to that of the surrounding "air," or changing the pressure trapped inside a gas bladder.

Organisms on gas giants are unlikely to be intelligent, at least in terms of what we Earthlings define as intelligent. On most gas giants there is no solid material from which to make tools. Interacting with your surroundings in any complicated way would not be necessary for survival. Rather, a creature that has evolved in a gaseous environment would likely be camouflaged like a jellyfish, as lack of hiding places would render them far too easy prey for predators.

This idea of thriving ecosystems on planets such as Jupiter or Bespin is by no means new. Back in the 1970s, Carl Sagan and Edwin Salpeter published a paper speculating about what such a gas giant ecosystem might look like. They dreamt up three main kinds of Jovian creatures. The first they dubbed sinkers, small organisms that were forever falling to their doom in the dense and hot lower atmosphere. And yet they lived long enough to produce tiny offspring that would be pushed into the relative safety of the upper atmosphere by the swirling Jovian currents. Sagan's other suggested Jovian creatures were floaters, similar to huge whales the size of cities, and hunters, sophisticated life-forms scouring the aerial reaches in search of prey.

THE WANDERING EARTH: ARE THERE ROGUE ALIEN WORLDS?

"In the Milky Way, it is perfectly commonplace for a weaker civilization to become the livestock of a stronger civilization. You will discover that being raised for food is a splendid life indeed. You will have no wants and will live happily to the end. Some civilizations have sought to become livestock, only to be turned down. That you should feel uncomfortable with the idea is entirely the fault of a most banal anthropocentrism."

—Liu Cixin, *The Wandering Earth* (2000)

THE WANDERING EARTH

The 2019 science fiction movie *The Wandering Earth* was described by *The Hollywood Reporter* as "China's first full-scale interstellar spectacular." It's the year 2061, and for no apparent reason, rather than waiting another five billion years to turn into a red giant, our Sun has decided to do exactly that, right now. In 2061. Naturally, such a huge solar expansion threatens to engulf the Earth's orbit within one hundred years. By the way, the actual existence of red giants was predicted in a science fiction story. In his famous 1895 novel *The Time Machine*, H. G. Wells told a prophetic and terrible tale about the end of the world. The tale's Time Traveler has journeyed to the end of time. The Solar System is in meltdown. The Earth is locked by tidal forces. And our planet is spiraling toward a red giant Sun, which hangs motionless in an endless Sunset.

Instead of waiting a mere thirty million years into the future for a red giant as Wells had fortuitously done, *The Wandering Earth* has the world's nations forced into a United Earth Government in a mere forty years' time. The Wandering Earth Project is initiated, which seeks to ambitiously migrate our planet out of the Solar System altogether and into the Alpha Centauri system. How? By using twelve thousand gargantuan fusion-powered engines built across the northern hemisphere, assisted by torque engines along the equator, to propel the entirety of planet Earth. A gravitational slingshot maneuver using Jupiter's gravity-assist helps propel and navigate the Earth along its interstellar journey.

The Wandering Earth's premise sounds ridiculous. And yet, plenty of the film's science holds up well. Once you've gotten over the near impossible idea of actually moving Earth in the first place, the rest of the ideas flow from that. You can see that the filmmakers did their homework, consulting with four scientists from the Chinese Academy of Scientists. But are there rogue alien planets in deep space, simply wandering the interstellar void? And, if so, how did they get there?

WONKY ORBITS

In a cosmos replete with surprises, there always seems to be room for random weirdness. In a Universe with such potential in such a vast collection of locations, all sorts of little oddities pop up. Maybe it's a planetary body with a very weird orbit. We have one in our own Solar System, of course. Pluto was famed for its wonky orbit, and perhaps that's one of the reasons Pluto's not a planet anymore. It was reclassified in 2006 as a Dwarf Planet. Its new name is 134340 Pluto. But it's still the tenth most massive body orbiting the Sun.

Wonky orbits can be of great scientific use. Consider the discovery of Neptune. In 1821, French astronomer Alexis Bouvard published orbital data on Uranus, Neptune's neighbor. Later observations showed sufficient deviations from the data so Bouvard suggested that another planet, or unknown body, was having a marked effect on the orbit of Uranus through gravitational interaction. After Neptune was then discovered in 1846, astronomers began looking for more distant planets. They noted the wonkiness in Neptune's orbit and thought the gravity of a farther planet might be to blame. And so the search for Planet X began. Even though Pluto was found in 1930, some astronomers think Planet X may be out there still.

ROGUE WORLDS

Maybe a planet's orbit around a star is just too weird for its own good. Maybe it's the star at fault, which has become strange and lost, roaming the cosmos at will. Or maybe it's a rogue world, which just sits between star systems. Pluto's not the only odd planet. Rogue planets (also known as interstellar, nomad, free-floating, unbound, orphan, wandering, starless, or Sunless planets; phew!) are interstellar objects of planetary mass, which have no host planetary system. Some may have been thrown out of their home system or have never been gravitationally bound to any star. Whichever it is, they are no longer tied to the gravity of their parent star, and now orbit the Galaxy directly. How many such planets exist?

Get this: the Milky Way alone may have billions to trillions of rogue planets. With numbers in space science being necessarily, well, astronomical, scientists hope soon to narrow the range of numbers used to estimate just how many rogue worlds there are. The upcoming Nancy Grace Roman Space Telescope has cutting-edge science objectives in cosmology and exoplanet research, which includes a census of exoplanets to help answer questions about the potential for alien life in the cosmos. Questions such as how common are Solar Systems like ours? What kinds of planets exist in the cold, outer regions of other planetary systems? And what factors determine the habitability of Earth-like worlds? The Roman Telescope uses techniques that will find exoplanets down to a mass only a few times that of the Moon, and so would be able to carry out a census sample of free-floating rogue worlds with masses likely down to the mass of Mars. But, as yet, only just over two dozen rogue planets have been discovered in the solar neighborhood.

WEIRD WORLD

Imagine being on a rogue planet. What kind of world would it be? Would it be warm enough for life? Well, there's no parent Sun, so there's no *external* source of heat. But research scientists have suggested that some rogue planets adrift in interstellar space might hold on to a thick atmosphere, which would not freeze out, and would be preserved by the pressure-induced qualities of a thick hydrogen-heavy atmosphere. Another positive factor in being a rogue planet is having no blazing Sun in its early years. As rogue planets got that way by being ejected from their planetary system probably during formation, they would get less of the star-made ultraviolet light that can strip away the lighter elements of its atmosphere. According to calculations, even a rogue planet the size of the Earth would have enough gravity to stop hydrogen and helium escaping its atmosphere.

Our imagined rogue world might also have oceans. In an Earth-sized object, the geothermal energy, which originates from the formation of the planet and from the radioactive decay of materials in the planet's core, could maintain a surface temperature above the melting point of water, allowing liquid-water seas to exist. Rogue planets are also likely to stay geologically active for long periods (and when geologists say long periods, believe it—that's usually billions of years).

So, a source of heat, a thick atmosphere, seas, and a stable environment are the kinds of conditions that might provide energy for life. If the rogue planet has, like Earth, a protective magnetosphere, and one which originates from the Earth's fluid outer core, then our imagined world may also have ocean-floor volcanism, and hydrothermal vents that could generate life-providing energy. These rogue worlds would be tricky to detect due to their weak heat emissions, but they would reflect stellar radiation and far-infrared thermal emissions, which may be detectable if less than one hundredth of a light-year from Earth. Roughly 5 percent of Earth-sized ejected planets with Moon-sized Moons would keep their natural satellites after ejection. And that satellite, of course, would be a major source of tidal heating.

ROGUE PLANET

Using this possibility of life on a rogue world, the *Star Trek: Enterprise* episode "Rogue Planet" speculated what life on such worlds might be like. A planet appears on the ship's sensors, adrift and without a solar system. An away-team ground crew is sent down to the planet to investigate. This small inhabited rogue world, thought to be ejected out of its orbit and traveling through interstellar space, exists in a state of perpetual night. A technologically advanced humanoid alien species known as the "Eska" are then encountered. The Eska have named the planet Dakala and explain that they have been visiting this rogue world for nine generations to hunt the wildlife.

One of the away-team hears a woman's voice calling his name. He finds a blonde woman in a clearing, but she runs off. Later, the woman is seen again, looking distressed, but again she disappears. Eventually, they encounter the woman, and she tells them that her kind are able to take on the essence of any life-form on the planet. Her species is distressed, as they are being hunted by the Eska and they want the hunting to stop. For their own part, the Eska explain that the reason they visit this rogue world is because the prey can sense their thoughts, which makes them an "admirable" challenge to hunt and kill.

The *Enterprise* crew find a solution by masking the chemical signature of the prey, and soon enough, down on Dakala, the Eska hunting party start to have problems detecting their prey. The Eska wonder why, but *Enterprise* personnel simply blame it on bad luck. After the Eska have departed Dakala, the blonde woman is encountered one last time. As she slopes off into the forest, she assumes the natural form of her kind—a large gastropod. So, yes, rogue worlds exist, and science fiction helps us imagine what life on such worlds might be like.

PART IV
ALIEN INVASION

ARE THE ANCIENT ALIENS IN *ASSASSIN'S CREED* POSSIBLE?

"It seems possible that the Earth has been visited by various Galactic civilizations many times, possibly ~10^4, during geological time. It is not out of the question that artifacts of these visits still exist, although none have been found to date, or even that some kind of base is maintained within the Solar System to provide continuity for successive expeditions."

—Carl Sagan and Iosif Shklovskii,
Intelligent Life in the Universe (1966)

ASSASSIN'S CREED'S ANCIENT ALIENS

The Assassin's Creed franchise of video games is ostensibly about the conflict between two sides in human history: the Templars and the Assassins. This conflict within the games, though played out in the shadows, is nonetheless one of history's main engines. It often involves the two sides' race to amass artifacts called "Pieces of Eden," which were left behind by the First Civilization millennia ago.

In the *Assassin's Creed* Universe, this First Civilization, the Isu, or sometimes called Those Who Came Before, were a highly advanced race who ruled over Earth roughly 77,000 years ago. Though they were humanoid and native to Earth, the precise origin of the Isu is unclear. And yet they were so advanced that they created homo sapiens to be their docile slaves, along with homo neanderthalensis to act as soldiers.

The *Assassin's Creed* Universe hypothesis of an ancient alien civilization is pretty typical. It's part of the "ancient astronauts" hypothesis that intelligent extraterrestrial beings visited Earth and made contact with humans in antiquity or prehistoric times. The *Assassin's Creed* backstory also echoes a popular theme in the ancient alien hypothesis: that gods or god-like beings from most of Earth's religions are, in fact, extraterrestrial in origin. It's also believed that ancient or early humans would have interpreted any advanced tech brought to Earth by ancient astronauts as evidence of divine status. After all, as Arthur C. Clarke said in his 1962 book *Profiles of the Future*, "any sufficiently advanced technology is indistinguishable from magic."

Though most academics and archaeologists consider the ancient aliens or ancient astronaut hypothesis to be unscientific, and this hypothesis has received no credible attention in establishment peer-reviewed studies, let's look at the possibility anyway. After all, if we can't do it in a book like this, while still remaining skeptical, which is of course the correct scientific position on *any* topic, where *can* we do it?

COSMIC TIME

Questions about ancient aliens first require us to think about two things: How old is the Universe, and from what part of the Universe might ancient aliens originate? Now, the Christian sweep of time was a mere bite-size history, one that began in Genesis and ended in Revelation. The Church scholars wanted to get a temporal grip on their history. How long ago had God created all this wonder? They began to add up the "begats," the long procession of scriptural births and deaths found in the Christian bible. The fashion for doing so started with Eusebius, Chairman of the Council of Nicaea in 325 AD. Eusebius claimed that 3184 years had elapsed between Adam and Abraham. Medieval German astronomer Johannes Kepler caught the dating bug. He estimated the date of Creation at about 3993 BC. Even world-famous physicist Isaac Newton followed suit. Newton put the date at 3998 BC.

For most of history, philosophers thought that the beginning of time was only about six thousand years ago. This turned out to be an immensely dumb idea, of course. Six thousand years is hardly any time for a planet to form and cool, for life to evolve, and for humans to get "smart."

It wasn't until the relatively modern discovery of fossils and rock dating that we humans discovered time was cosmic. Huge, in fact. Long story short: planet Earth turned out to be 4.54 billion years old. And geologists and physicists estimate that the Universe itself is about 13.77 billion years old. Since life in the Universe has had all that time to develop, it's a fair and almost entirely contemporary question to ask: When might alien life have begun? This question is at the heart of the backstory in *Assassin's Creed*.

That strange breed of humans who call themselves astrophysicists believe time began with a so-called Big Bang. Around 13.77 billion years ago, they say, space, time, and everything else was created out of nothing. The early Universe was hot, but cooled down to form atoms, the building blocks of matter. These atoms formed the simple elements of hydrogen and helium. The heavier elements were cooked up later, by stars. That much is simple enough. But fundamental elements have to coalesce into complex planetary systems before life can evolve (as far as we know, anyway). So, when did the first planets evolve?

ARE ROCKY PLANETS LATE STARTERS?

Some scientists think that the cosmos needs two or three generations of stars before it can generate a planet like Earth. It takes that long, they say, to make the elements out of which rocky planets like Earth are made. Other scientists claim that rocky planets may have been made with the very first stars.

For example, consider a report in the UK's *Daily Telegraph* newspaper from September 2021 claiming that alien life in our Milky Way Galaxy may be far more likely than first thought. And the reason for this claim was the discovery that the basic chemical conditions which resulted in life on Earth could exist more widely across the Milky Way. Scientists found significant amounts of large organic molecules surrounding young stars. A research team studied the discs of swirling material which surround stars and eventually come together to form planets. They found large reservoirs of precursor molecules which are stepping-stones to the complex molecules needed for life, such as sugars, amino acids, and ribonucleic acid. Each molecule emits light at distinctly different wavelengths, producing a unique spectral "fingerprint" which can be picked up by the Alma telescope in Chile. (Alma has enabled astronomers to look for such molecules in the innermost regions of these disks, on size scales similar to our Solar System, for the first time.) Analysis shows that the molecules are primarily located in these inner disc regions with abundances between ten and one hundred times higher than models had predicted, so it's possible that the molecules needed to kick-start life on planets are readily available in all planet-forming environments.

This evidence is local and recent and doesn't necessarily speak the same language as that of deep time. That's because around 400,000 years after the Big Bang, the cosmos was a cold and dark murk of only hydrogen and helium but, 400 million years later, it started to shine with the light of newborn Galaxies. So, sometime in between, the first stars must have kick-started into life. What were they made of? How bright and how big were they? How long were their lives, and what happened to them after their deaths?

No one knows. In truth, no scientist can truly say with any precision what the first stars were like. And none of today's powerful 'scopes, whether it's Hubble, Spitzer, or Keck, have been able to find them. But astronomers *do* have some ballpark clues, of course. Those first stars, known as Population III stars, were in all probability made almost entirely of hydrogen and helium, the elements formed as a result of the Big Bang. These first-generation stellar children would not have had in their makeup heavier elements such as the crucial carbon, nitrogen, oxygen, and iron that are found in today's stars like our Sun and are considered vital for life. In short, Population III stars were metal-free (a brief reminder here that astronomers call any element heavier than helium a metal).

Naturally, given that no one has actually observed any metal-free stars, you might feel the above statement to be overly bold, yet from observations, experiments, and calculations we are pretty confident in our theory that only hydrogen and helium and a mere dash of lithium were formed directly after the Big Bang. The only way that heavier elements like carbon, nitrogen, oxygen, and iron can form is by fusion of lighter elements in the cores of stars. So, until the first stars began to form them, none of these elements existed in the Universe. The first stars must have been metal-free.

To some extent, this argument is a tad academic. For us, it means that the possibilities are that an alien civilization could either be ancient, or *really* ancient. If the consensus is correct, that the cosmos needs two or three generations of stars before it can generate a planet like Earth, then an ancient civilization might be, say, three billion years old. If the minority is right, that rocky planets may have been made with the very first stars, then an ancient civilization might be as old as ten billion years. Such ancient civilizations are usually taken to mean a civilization which is wise, sophisticated, and technologically advanced. Luckily for us, astrophysicists have already considered types of alien civilization, so let's look at what's gone before.

WHAT KIND OF ALIEN CIVILIZATION? THE KARDASHEV KINDS

Back in 1964, the Soviet astrophysicist Nikolai Kardashev suggested three levels of alien civilizations. These levels were based on what Kardashev thought might be the different orders of magnitude of power that may be available to sophisticated civilizations. So, Kardashev's scale has three designated categories. Type I civilizations, also called planetary civilizations, would be able to use and store all of the energy available on their home world. Type II civilizations, also known as stellar civilizations, would be able to use and control energy at the scale of their stellar system. And Type III civilizations, also called galactic civilizations, would be able to use and control energy at the scale of their entire host Galaxy.

We humans have not yet reached Type 1 civilization status. We are not yet able to use and store all of the energy available on Earth. In fact, quite the contrary. Our energy use is *out* of control, given global warming. But what type of Kardashev civilization might we associate with the ancient aliens or ancient astronaut hypothesis? It would seem unlikely that a sophisticated spacefaring civilization would be a mere Type I civilization, which had just gotten its collective limb onto the lower rung of the cosmic ladder. And since it would seem that an entire Galaxy lies at the alien feet of a Type III civilization, it's not clear why they would come and visit a poxy planet like Earth. So perhaps we are looking for a Type II civilization, one that was able to use and control energy at the scale of their stellar system, such as building a Dyson Sphere, and is now looking to branch out into other stellar systems.

WHAT KIND OF ALIEN GALAXY?

What kind of Galaxy is most likely to have produced such a civilization? Let's briefly remind ourselves of the different types of Galaxies. As the basic building blocks of the cosmos, some Galaxies are simple while others are quite complex in nature. American astronomer Edwin Hubble came up with a classification scheme of Galaxies way back in 1926. It was the first step toward a theory of Galaxy evolution.

Though now considered an oversimplification, the basic ideas of Hubble's so-called tuning fork diagram still hold. As you can see, Hubble's diagram is divided into two parts: elliptical Galaxies (ellipticals) and spiral Galaxies (spirals). He also numbered the ellipticals from zero to seven. These characterize the ellipticity of each Galaxy. For example, E0 is nearly round, while E7 is very elliptical. Likewise, the spiral Galaxies were given letters from "a" to "c" to help characterize the compactness of the Galaxy's spiral arms. For instance, an "Sa"

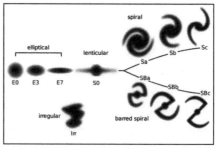

Hubble classification scheme

spiral is tightly wound, whereas an "Sc" spiral more loosely so. It's also worth keeping in mind the size of the so-called bulges in spirals—the rounded central regions—which increase in size the more tightly the spiral arms are wound.

The classification is not exact. For example, there's a close connection between the bulges of some Galaxies (types S0, Sa, and Sb) and elliptical Galaxies. They may very well be similar objects. For some time, the Hubble tuning fork was considered an evolutionary sequence. It was thought that

Galaxies evolved from one type to another, progressing from left to right across the diagram. But, once more, the initial idea was found to be an oversimplification. Nonetheless, astronomers still use this classification today, and it will help with our discussion of the ebb and flow of internal star birth of Galaxies.

So, what classification of Galaxy is most likely to have produced an ancient civilization? Well, recent studies suggest that the cosmos may have good and bad neighborhoods. And life will probably evolve best in huge elliptical Galaxies, whereas so-called dwarf Galaxies are considered the least hospitable. Spiral Galaxies like our own Milky Way fall somewhere between the two extremes. Naturally, this idea that the cosmos has more or less life-giving regions is still suppositional, especially as human scientists still haven't found any examples of alien life. But, as we have seen, habitable zones in which water should be stable and Earth-like creatures have a good chance of survival have been suggested for alien solar systems and regions within Galaxies.

BEWARE STARS THAT GO BANG

Very recent studies have gone one step further in trying to identify the most habitable types of Galaxies. Astronomers compare different Galaxies to the only known example of an inhabited Galaxy: our Milky Way. Astronomers believe that life-friendly Galaxies need lots of stars that can host planets, but a low rate of star formation to limit the number of supernovae. Why? Because the violent explosions of supernovae, which happen when massive stars die every few million years or so, may lead to mass extinctions on thriving worlds in the local stellar neighborhood. A relatively low rate of recent star formation would suggest fewer stars that are ticking time bombs, and a more quiescent region.

Studies show that the clear galactic winners are giant ellipticals, which are more than twice the mass of our Milky Way, but have less than a tenth the number of volatile young stars. The math shows that if the Milky Way has a capacity of hosting one habitable planet, the giant ellipticals would host as many as 10,000 habitable planets, based on the known supernova rate in our Galaxy.

And the *worst* galactic places to find alien life in the cosmos? Those small, irregular Galaxies, labeled "Irr" on our diagram above, which have lots of newborn stars within them. In irregulars, consistent and nagging supernova blasts would sterilize whole galactic neighborhoods. And there are, in all likelihood, insufficient heavier elements with which to form planets.

GALAXIES DANGEREUSES

Supernovae aren't the only galactic danger to alien life. There's also Gamma ray bursts. Though rarer, these are often deadlier explosions than their supernova sisters. Earth is thought to have been hit by only one lone gamma ray burst in the last half a billion years. And *that* possibly caused the Ordovician mass extinction that so terrorized the poor trilobites around 440 million years ago. And yet, an Earth-like world in an active dwarf Galaxy would have to weather one hundred such gamma ray bursts during the same time window.

As with supernovae, gamma ray bursts are more common in regions of star formation. And astronomers believe that gamma ray bursts reached their peak about eleven billion years ago. That's when star formation was in its heyday across the cosmos, and a scarcity of heavy elements made gamma bursts a more likely result of star explosions. In those days, almost all Galaxies were dangerous, no matter their mass.

What does all this mean for the age of ancient alien civilizations? It means the so-called cosmological habitable zone extends in time, as well as in space. In other words, advanced life would be delayed by blasts of extreme radiation for the first few billion years of the cosmos. Do not be fooled by today's Universe. Today's calm contrasts wildly with yesterday's chaotic cosmos. So, if we are trying to paint a picture of the timescale of an ancient alien civilization, it seems that the science suggests somewhere between three and ten billion years old should do it.

FOUNDATION AND *MANDALORIAN*: HOW MIGHT ALIEN EMPIRES GROW?

"Space opera: A popular item of science fiction terminology, echoing the practice (dating from the 1920s) of referring to Westerns as 'horse operas,' and more immediately the term 'soap operas' (from 1938) for never-ending radio series: when radio was the principal medium of home entertainment in the USA, daytime serials intended for housewives were often sponsored by soap-powder companies, and hence the nickname. 'Soap opera' was quickly generalized to refer to any corny domestic drama. The pattern was extended into science fiction nomenclature by Wilson Tucker, who in 1941 proposed 'space opera' as the appropriate term for the 'hacky, grinding, stinking, outworn, spaceship yarn.' It soon came to be applied instead to colorful action-adventure stories of interplanetary or interstellar conflict."

—John Clute and Peter Nicholls,
The Encyclopedia of Science Fiction (2011)

"Human history is like a river. From any given vantage point, a river looks much the same day after day. But actually it is constantly flowing and changing, crumbling its banks, widening and deepening its channel. The water seen one day is never the same as that seen the next. Some of it is constantly being evaporated and drawn up, to return as rain. From year to year these changes may be scarcely perceptible. But one day, when the banks are thoroughly weakened and the rains long and heavy, the river floods, bursts its banks, and may take a new course."

—Hubert Kay, *Life* (1948)

SPACE OPERA

Sleek starships and big space battles. Dashing heroes and camp droids. Evil galactic empires and exotically colored "space babes." Such are the tropes of space opera. Frequently set in a spacefaring civilization, usually framed in the far future or a long time ago, in a Galaxy far, far away. The tech is everywhere, but always secondary to story. Space opera has an epic sheen on it: the cosmos is huge, its many civilizations sprawling, and the empire rife with conflict and intrigue. It's *Game of Thrones* in space. (Actually, maybe *Game of Thrones* is space opera on terra firma!) In space opera, the plot reaches at least across a stellar system, and often deep into the Galaxy and beyond.

In Isaac Asimov's legendary *Foundation* trilogy, now also an Apple TV drama series, the unlikely hero of the space opera is a math professor named Hari Seldon, who predicts the future using what Asimov calls "psychohistory." Math is used to model the past to help anticipate what will happen next, including the fall of the inevitable galactic empire. And in *The Mandalorian*, an American space opera TV series created for the streaming service Disney+, the first live-action drama in the Star Wars franchise has a plot that follows Din Djarin, a lone Mandalorian bounty hunter in the outer reaches of the Galaxy.

One can assume that the authors of space opera figure that, if we've had empires on Earth, why not also empires in space? After all, when the New World was discovered by ship in medieval times, writers began to imagine alien worlds in space (based on the logic that if new "worlds" can be found on Earth, why not also find new worlds in space?). But how might empires like that in *Foundation* and *The Mandalorian* grow? What science would we need to consider before the chance associations of evolution and development result in the lucky outcome of a galactic empire?

SPEEDS OF LIFE

Imagine you're a planet maker, a little like Slartibartfast, the Magrathean designer of planets in *The Hitchhiker's Guide to the Galaxy*. Your planet sits warmly and happily in the Goldilocks Zone, orbiting its parent star. Your created world has all the common ingredients for life, or at least the ingredients that seem to work on Earth, according to current science. Will the same ingredients work on other worlds? Does life's pathway always move at the same rate, or is it possible to speed up life, and enjoy evolution in the fast lane?

It's an interesting question. We've seen that life's pathway, evolution, depends on many things. From bacteria to blue whale, size is certainly a factor. Changes in the environment, such as comet strikes and climate changes, will also affect the speed of life. Scientists are not certain how many species roam the Earth. Estimates vary from two million to one hundred million species. If we ignore bacteria, our best guess is about ten to fourteen million species. But be warned, we're still counting, adding thousands of new species each year.

The time scale for evolutionary change is very long. A typical period for the appearance of one advanced species from another is about 100,000 years. Scientists believe that humans branched off from their common ancestor with chimpanzees about five to seven million years ago. In that time, human subspecies, like Neanderthals, developed and died. Consider also bacteria. They occupy a wide variety of habitats. They have a broader range of chemistries than any other group. They are adaptable, indestructible, and astoundingly diverse. And their speed of life has been long and ancient.

Are there speedier types of life out there in space? Well, there could be. Evolution depends partly on the way our genetics react to change. The rate of evolution on another world could be speedier if alien genes changed more quickly and didn't just wait for a random gene mutation to force evolution (assuming aliens *have* genes, of course).

EVOLUTION AND EMPIRE

Imagine that one planetary civilization becomes dominant in a particular region of space, such as Coruscant in the *Star Wars* Galaxy, which sits among the populous Deep Core worlds, or the planet which is the creative basis for the *Foundation* series, Trantor, which is in some senses based on Rome at the height of the Roman Empire.

In academic economic geography, we humans have a key concept known as the law of uneven and combined development. This economic law attempts to explain why Eurasian civilizations have survived and conquered others, while arguing against racist ideas that Eurasian hegemony is due to any form of Eurasian intellectual or genetic superiority.

Might there be some kind of law of uneven and combined development for empires in space? This unwritten law could be applied to describe the dynamics of history and account for the interaction of a planetary imperial development whose world units are highly heterogeneous. In other words, the law would describe why some planets develop at a faster rate and overcome those less developed. To get a grasp of how empires might grow in space, we could think about the development of empires on Earth.

Describing the development of terrestrial empires is no easy matter. Our theory would have to take in the Roman Empire, the Macedonian Empire of Alexander the Great, the Mongol Empire, the dynastic Iberian Union when Spain and Portugal briefly ruled as one, the Empire of Japan, the Ottoman Empire, the many Chinese empires, and the French, Spanish, and British Empires, the last of which ruled over one quarter of the globe. Our theory would also have to see history not as a mere series of isolated facts but rather as a general process, whose laws govern the evolution of human society. We need to look beyond the particular and arrive at the general, just as Hari Seldon does in Asimov's *Foundation*.

In the discussion that follows on the dynamics of history, it would be wise to wear two thinking caps: one regarding different countries within a single world like Earth, and another regarding different planets within a potential empire.

THE ENGINE OF EMPIRE

What fundamental motor drives the stages of history and their transitions from one social system to another? In short, the answer is trade and economy. Particularly with the influence of science and tech from the sixteenth century onward, the characteristic evolutionary pattern of worldwide expansion has been through the growth of a world economy. An economy that linked up more and more peoples and lands together through trade, migration, and investment.

Human history has been the development of humanity's endeavor to raise itself above the natural animal level. Our "mission" began millions of years back when our distant humanoid ancestors first walked upright and freed their hands for manual work. Ever since, consecutive phases of social evolution have developed on the basis of changes in our power over nature. Specifically, human society has progressed through a series of stages based on different modes of material production associated with human power and ingenuity.

If we examine the process of human history, all the way back to prehistory, we are struck by the remarkable slowness with which our species evolved. Our gradual evolution away from the bestial toward a genuinely human condition happened over millions of years. The first crucial step was our separation from our simian ancestors. This development was, naturally, blind, as the evolutionary process is without an objective or specific goal. But our hominid ancestors, initially by walking upright, next by using and then producing their tools, found a niche in the environment which impelled us forward.

Descartes was wrong. It's not "I think, therefore I am." It's "I am, therefore I think." The development of human society has progressed through a series of stages, from hunting and gathering, through pastoralism and cultivation, to today's commercial society. From a scientific point of view, material progress is primary. Ideas, thought, and consciousness are secondary. In other words, consciousness and human ideas about the cosmos come out of our material conditions rather than the other way around. In short, before men and women can develop art, science, religion, and philosophy, they must first have food to eat, clothes to wear, and houses to live in. I am, therefore I think.

ALIEN EMPIRES

On each alien planet within a potential empire, we can assume that similar processes of history to that described above also emerge. We can assume that, despite trade, even galactic trade, the different planets develop and progress to a large extent *independently* of each other. Not only are the planets *quantitatively* unequal, in terms of local economic and population growth, but also *qualitatively* different, in terms of their particular planetary cultures and geographical features. That is, planets developed their own specific history with global peculiarities.

On the other hand, the numerous planets did not exist in total isolation from one another. They evolved as integral parts of a galactic whole, a cosmic society, a larger totality. They all coexisted together. As a result, they also shared plenty of characteristics. And they influenced one another through cultural diffusion, such as Buddhism and Christianity on Earth, trade, such as our Silk Road, and political relations and other "spillover effects" from one planet to another.

Historically, this would have a number of effects. Less progressed, older, or more primitive planets would adopt parts of the culture of more progressed or modern planets. And a more progressed planetary culture might also adopt or merge with parts of a more primitive culture, with good or bad results. This would mean that cultural practices, institutions, traditions, and ways of life belonging to both very old and very new epochs and phases of galactic history would combine, juxtapose, and link together, within one planet. And *that* would mean one could not really say that different planetary societies all developed simply through the same sort of linear sequence of necessary developmental stages. Instead, planets would adopt the results of developments already reached elsewhere, as has happened on Earth between cultures, without going through all the previous evolutionary stages which led up to those results. Some planets would skip stages which other planets took millennia to go through, or very rapidly modernize, which would have taken other planets centuries to achieve.

Because of this planetary coexistence, planets could both *aid* or *advance* the progress of other planets through trade and resources. On the other hand, they could also *block* or *brake* other planets as competitors, halting their progress by preventing the use of capital, tech, trade, labor, and land. And so, the domination of one planet by another doesn't mean that the dominated planet is *prevented* from progressing at all. Instead, it develops according to the needs of the dominating planet. For example, one can imagine a dominated planet with huge resources of iron, perfect for making a Death Star. An export industry develops around mining that iron in the dominated planet, but the rest of the planet's economy is not developed. The planet's economy becomes unevenly developed and unbalanced. (An example of this trade domination can be found in the history of the British Empire in India. In an attempt to break the Chinese monopoly on tea, the British bypassed China by using their imperial position to enforce tea plantations in India.) Another example would be education. A school system is set up on a dominated planet with "alien" assistance, but the schools teach only the messages that the dominating planet wants taught.

The major trends and tendencies happening at galactic level could also be found on each separate planet, where they might fuse with distinctive local planetary trends. And yet this locally specific mix would be varied. Some galactic trends assert themselves stronger or faster, others weaker and slower on each specific planet. So, like countries within an empire on Earth, a planet could be very progressed in some activities, but relatively dysfunctional in others. Thus, the planetary response to the same events of galactic significance could be quite different on different planets, as the local people attached different gravitas to cultural experiences and drew different conclusions.

There are many scientific ways in which we can imagine the growth of empires like that in *Foundation* and *The Mandalorian*. All it takes is multiple living planets with thriving civilizations at different stages of development. So, maybe it's not so fanciful when authors of space opera imagine empires in space based on the imperial past of Earth.

GROOT, SPOCK, OR BUG: WHAT MIGHT ALIENS LOOK LIKE?

"If our Solar System is not unusual, then there are so many planets in the Universe that, for example, they outnumber the sum of all sounds and words ever uttered by every human who has ever lived. To declare that Earth must be the only planet with life in the Universe would be inexcusably bigheaded of us . . . If an alien lands on your front lawn and extends an appendage as a gesture of greeting, before you get friendly, toss it an eight ball. If the appendage explodes, then the alien was probably made of antimatter. If not, then you can proceed to take it to your leader."
—Neil deGrasse Tyson, *Death by Black Hole* (2007)

GROOT, SPOCK, OR BUG?

What might we find on alien worlds? Groot, the extraterrestrial, sentient, tree-like creature from the Marvel Universe? Spock, the half-human, half-Vulcan, uber-rational alien from the planet Vulcan? The Bug, the giant, scavenging cockroach from the stars who features in *Men in Black* and is stuffed into the skin of a human named Edgar? All pretty weird, yet aliens could get a lot weirder. Consider the potential for extraterrestrial weirdness by looking at some examples of unusual aspects of life on Earth.

What super senses might the alien have? Just think of the super senses of some terrestrial creatures. A vulture can spot a carcass from a great distance, the four-eyed fish can see above and below water simultaneously, a fly's multifaceted eye sees a very different world than a human eye, while other insects can see into ultraviolet light. And lions have an area on the retina that actually empathizes with their prey.

The home planet of the alien will also be important. On Earth, courting, egg-laying, hibernation, and other rhythms that govern life are influenced by the cycles of the Earth, Moon, and Sun. Every animal's perception of time varies according to its heart rate. A shrew lives thirty times faster than an elephant, so time appears to pass more slowly. Each creature has a unique view of the world made up using a combination of different senses. The mind creates mental maps for navigational skills, which can also be affected by genetic programming. Other super senses have resulted from the need to hunt or avoid becoming a meal. The mind decides on skills it needs to survive. So it is on Earth, so we can assume it would be so on other worlds.

What curious alien idiosyncrasies might extraterrestrial possess? Terrestrial animals use senses of which humans are unaware. Sensitivity to the Earth's electromagnetic fields or to weather pressure can be used to aid navigation. Some animals can predict Earthquakes. Predators put these senses to lethal use: a shark homes in on the body electricity of its prey. Smell is invaluable in hunting, protecting a species, mating, and navigation. Petrels use it to find fish in the open sea, springboks emit an "alarm" odor to warn the herd of a predator, salamanders inject their females with aphrodisiac, and a salmon's epic journey across the ocean to spawn and die is achieved through its sense of smell.

Our ears have a limited range and are deaf to a low-register elephant conversation or the high-pitched squeaking of mice. Whales use sonar to communicate across hundreds of miles of sea or to locate prey while spiders listen for the wing beats of prey and the kangaroo rat has sensitive hearing. Would aliens have these super senses too? Of course, some super senses may also be unseen, such as telepathic ability, eyesight, hearing, and super intelligence that film and fiction so often associate with extraterrestrials.

WHAT TERRESTRIAL ANIMALS REVEAL ABOUT ALIENS

For as long as we've known Earth is but a mere planet in a vast and unending cosmos, we humans have wondered about the mystery of alien life. Even major scientists have had a ponder on this profound question. Pioneering astronomers, such as the father of planetary motion, Johannes Kepler, and the discoverer of Uranus, William Herschel, both believed in the existence of extraterrestrial life. Stargazing through his ground-breaking telescopes, Herschel imagined he spied towns and forests on the surface of the Moon. And yet science is still no closer to imagining what extraterrestrials might look like.

Perhaps life on Earth provides some clues as to what life on other worlds might be like. Maybe there are universal laws of biology that not only govern terrestrial life, but also apply to aliens. Of course, chief among these laws is that species evolve by natural selection. This is the foundational idea of evolutionary biology proposed by Alfred Russel Wallace and Charles Darwin. Irrespective of the way in which extraterrestrial biochemistry works, and quite apart from how planetary environments might vary, some sort of natural selection would be at play and could have corralled alien evolution to limited menus of options.

Alien life may bear striking parallels to Earthly life. Certainly, this is the hope of so many writers and moviemakers who take rather wild guesses when they represent extraterrestrials in film and fiction. But to be fair, it could well be that most aliens will be bilaterally symmetrical (the body plans of most animals, including us, exhibit this mirror symmetry, as they are symmetric about a plane running from head to tail, or toe) and use familiar forms of locomotion (such as limbs, paddles, or jets). Do aliens have sex? Sadly, and according to the same writers and moviemakers, aliens appear to have kept this private, along with much else.

AS WITH HUMANS, SO WITH ALIENS

Most scientists avoid the fictional game of suggesting any particular vision of what extraterrestrials might look like, so no Groot, Spock, or Bug, and no Wookiees, Ewoks or BEMs. Rather, scientists concentrate on how aliens might *behave*. It's a fair guess that at least some aliens will exhibit social cooperation, tech, and language. And in all probability, extraterrestrials will share the quality we humans hold most precious: intelligence. When you look at the legions of smart extraterrestrials in film and fiction, it seems inevitable that aliens will be smarter than us.

Indeed, it's even possible that the alien's backstory might mirror ours. They too might cherish their own "becoming" myth of a social and intelligent creature, with the art of language, which has developed complex tech based on sound scientific concepts. It is difficult to see how any other backstory would explain a star-faring species. If you're building starships and sailing the cosmos without first destroying yourselves, you can't be dumb.

Like us humans, advanced star-faring extraterrestrials living on alien worlds will also be the culmination of billions of years of evolution before their species appeared. Like us, they will be latecomers to the long evolutionary drama on their home world, just one lineage among billions of species. And like us, they are also likely to be evolutionary oddballs: big-brained creatures with language, science, and complex tech with the ability to alter their planetary habitat as a base for exploring other planets beyond their home world.

Since the big brains of humans come equipped with big imaginations, we often invoke superior alien intelligence in our own origin story. Maybe, as in Ridley Scott's *Prometheus* and the *Star Trek: Next Generation* episode "The Chase," "alien seeders" gave us humans life. That would make us Earthlings an experiment conducted by a superior intelligence, just as Douglas Adams suggested in *The Hitchhiker's Guide to the Galaxy* where superintelligent white mice designed the Earth to find an answer to the "Ultimate Question" of the meaning of "life, the Universe, and everything."

The main point about all this speculation as to what aliens might look like or how they might behave is that it's fine to speculate as wildly as you want, as the chances of your hypothesis being tested in your lifetime through the discovery of intelligent aliens is so remote as to be almost entirely disregarded. Writers, moviemakers, and even some scientists can speculate until the cosmic cows (don't) come home.

TIME LORDS, VULCANS, AND KRYPTONIANS: ARE CLEVER ALIENS VIOLENT?

"The surest sign that intelligent life exists elsewhere in the Universe is that it has never tried to contact us."
—Bill Watterson, *The Indispensable Calvin and Hobbes* (1992)

"It [the beauty of physics] has nothing to do with human beings. Somewhere, in some other planet orbiting some very distant star, maybe another Galaxy, there could well be entities that are at least as intelligent as we are, and are interested in science. It is not impossible. I think there probably are lots. Very likely none is close enough to interact with us, but they could be out there very easily. And suppose they have [. . .] very different sensory apparatus and so on, they have seven tentacles, and they have fourteen little funny-looking compound eyes, and a brain shaped like a pretzel, would they really have different laws? Lots of people would believe that, and I think it is utter baloney. . . . They [the three principles in nature] are emergent properties. . . . Life can

emerge from physics and chemistry plus a lot of accidents. The human mind can arise from neurobiology and a lot of accidents."
—Murray Gell-Mann, *Beauty and Truth in Physics* (2007)

BEWARE THE ALIEN

The famed British physicist Stephen Hawking seemed to have a love-hate relationship with aliens. On the one hand, Hawking helped launch major efforts in the search for signs of intelligent alien life in the cosmos. But, on the other hand, he thought it likely that such creatures would try to destroy humanity. For the best part of the last decade of his life, Hawking spoke publicly about his fears that an intelligent alien civilization would have no problem wiping out humanity the way a human might wipe out a colony of ants, for example. The late American evolutionist Stephen Jay Gould seemed to agree with Hawking. Gould pointed out that intelligent life may be commonplace but not long-lived: "Perhaps any society that could build a technology for such interplanetary travel must first pass through a period of potential destruction where technological capacity outstrips social or moral restraint. Perhaps, no, or very few, societies can ever emerge intact from such a crucial episode."

What was the basis of Hawking's conclusion? Well, intelligence. After all, it was meant to be relatively *intelligent* humans that have a hateful history of maltreating, and even massacring, other human cultures that are less technologically advanced and "intelligent." As quoted by the BBC in 2010, Hawking gave one of history's most infamous examples: "If aliens visit us, the outcome would be much as when Columbus landed in America, which didn't turn out well for the Native Americans." Why, Hawking asks, would an alien civilization be any different?

It seems Hawking's desire to know if there's intelligent alien life in the cosmos trumped his fears. In 2015, Hawking was part of a public announcement for a new initiative called *Breakthrough Listen,* a $100 million project funding thousands of hours of dedicated telescope time on state-of-the-art facilities in the most comprehensive search for alien communications to date.

The *Breakthrough Listen* project searches for signs of intelligent life. It doesn't broadcast signals from Earth. Indeed, Hawking voiced his fears at the *Breakthrough* launch event, saying "We don't know much about aliens, but we know about humans. If you look at history, contact between humans and less intelligent organisms have often been disastrous from their point of view, and encounters between civilizations with advanced versus primitive technologies have gone badly for the less advanced. A civilization reading one of our messages could be billions of years ahead of us. If so, they will be vastly more powerful, and may not see us as any more valuable than we see bacteria." And Hawking isn't the *only* scientist who has voiced concern over hailing the attention of intelligent alien civilizations.

Having said all that, *Breakthrough Message*, an associated program, studies the ethics of beaming messages out into deep space for intelligent aliens to pick up. Part of this second initiative is an open competition with a US $1 million prize pool to design a digital message that could be beamed out from Earth to a potential alien civilization. The program does, however, assert "not to transmit any message until there has been a global debate at high levels of science and politics on the risks and rewards of contacting advanced civilizations."

TIME LORDS, VULCANS, AND KRYPTONIANS

As we have said before, scientists still have no idea what alien life-forms might look or behave like. And that means we also have no idea how they might respond to contact from humans. In a 2010 Discovery Channel television show called *Into the Universe with Stephen Hawking*, Hawking said "Advanced aliens would perhaps become nomads, looking to conquer and colonize whatever planets they could reach. If so, it makes sense for them to exploit each new planet for material to build more spaceships so they could move on. Who knows what the limits would be?" Indeed, they may already know we're here.

Science fiction is, of course, replete with intelligent aliens. Take the Time Lords, the humanoid inhabitants of the planet Gallifrey in *Doctor Who*. Most famous for the creation and attempted monopolization of time-travel tech and their nonlinear perception of time, the Time Lords communicated by telepathy and could link their considerable minds to share information and further enhance their powers. Though described as highly intelligent, pompous, and "dusty" by other races, the Time Lords were also prone to small bouts of evil, such as putting the Doctor on trial, exiling the Doctor to Earth, trapping the Doctor in his confession dial, and so on.

Then there's the Vulcans. A fictional alien humanoid species in the Star Trek franchise's various television series and movies, the Vulcans are noted for their attempt to live by logic and reason, and with minimum interference from emotion. In the *Star Trek* Universe, they were the first alien species to make contact with humans. Have the Vulcans ever been villains? Well, it *could* be argued that Vulcans are so convinced of their own intellectual superiority that they often end up being hateful, prideful, and malicious. And, even though they are founding members of the Federation, Vulcans are secretly villainous, as devious and duplicitous as Romulans, but all the more Machiavellian because of their status as allies. But they're hardly genocidal villains.

Meanwhile, the Kryptonians are another fictional race of humanoid aliens, this time within the *DC Comics* Universe. As the dominant species of the planet Krypton, Kryptonians are hard at first to distinguish from Terran humans if we go on their appearance and physiology alone. But, looking under the hood, as it were, at their genetics, we find a very different tale. Their DNA is so complex that terrestrial science is not advanced enough to decode their genome. And their cellular structure allows for solar energy to be absorbed at such high levels that, when one of them comes to Earth and soaks up our yellow Sun's light compared to Krypton's red supergiant parent star, they get a vastly higher energy absorption. This absorption leads to those superhuman powers, including superhuman strength, speed, flight (obviously!), and other superhuman senses, though intelligence is not specially singled out.

Having taken a look at three representative intelligent alien races from science fiction, does it follow that Stephen Hawking is right about such aliens? Would they *really* have no problem wiping out humanity the way a human might wipe out a colony of ants? Let's first look at the general idea of intelligence and evolution.

INTELLIGENCE AND EVOLUTION

Is there an evolutionary advantage to being intelligent? And what does it mean to be "intelligent" anyway? We think intelligence is the one thing that separates us from all other animals, but don't other creatures show degrees of intelligence too? Is intelligence even necessary for evolutionary progress? Finally, given all these questions, under what conditions do some species become more intelligent than others?

It's not a simple matter, trying to compare human intelligence to the intelligence of other creatures. Animals can't read or write, so it's a bit of a challenge measuring just how smart they are. And yet in 2021, a study of six "genius dogs" progressed our understanding of dogs' memories, suggesting some of them possess an incredible grasp of human language. Hungarian scientists spent over two years combing the globe for canines who could recognize the names of various toys. Now, most dogs can learn commands to some extent, but learning the names of items appears to be a very different deal, and most dogs don't get to master this skill.

But six dogs, all border collies—herding dogs originally bred to work sheep—made the cut after showing they'd learned the names of more than twenty-eight toys, with some dogs knowing more than one hundred. The gifted dogs learned new names of toys with an amazing speed, remembering a new toy name after hearing it only four times. This canine rate of learning is comparable with that of human infants at the start of their vocabulary spurt when they suddenly start stringing words together at about eighteen months of age. The smarter dogs could also recall the names of the toys when they were tested again months later. Nor is this memory and intelligence talent particular to border collies. Dogs from other breeds also made the mark, including a German shepherd, a Pekingese, a mini-Australian shepherd, and a few dogs of mixed breeds. By studies such as this, scientists hope to better understand the relationship between animals and intelligence.

One thing we *can* say already, however, is that intelligence has cropped up many times in Earth's tree of life, and on numerous branches too. Mantis shrimps, octopuses, whales, dolphins, and the apes, among many others, show degrees of intelligence. Evolution seems to produce smart creatures.

But are all successful species intelligent? It would seem not. Consider bacteria. They're not exactly smart or intelligent in the usual sense. And yet they've been remarkably successful on Earth. In deep time we have plenty of other examples of successful organisms that aren't in the slightest bit intelligent. To us it seems that humans are by far the most successful species. We have a unique intelligence, along with a scientific and technical capacity. And we tend to gauge potential extraterrestrial intelligence against this unique intelligence. But it's worth bearing in mind that how you see the world depends upon the kind of questions you ask.

To get some kind of clues as to how aliens might get so clever, it's worth asking how humans got so smart. All species have to compete for space and food. But evolution is by no means all about competition. Humans thrived through cooperation too. Early humans developed skills such as the use of tools. But the key development may have been the way we interacted with one another. We learned how to communicate. And this would have been an essential step on the road to "Team Human."

Much has been written about brain size too. But it's not just the *size* of the brain. Brain size usually increases with body size in animals. Larger animals have larger brains, as they need them to control their bodies. It's also a question of what you actually do with your brain. The size of a modern human's brain is like that of an ancient human, but the cultural difference is huge.

DEGRASSE TYSON ON HAWKING

So, what's the answer to Professor Hawking's contention that clever aliens are necessarily violent? One answer to this question was provided by American astrophysicist and science communicator Neil deGrasse Tyson. In the Joe Rogan podcast #919, recorded in February 2017, deGrasse Tyson said, "Okay, [Hawking] is worried about the possibility of aliens enslaving us, based on the reality that we've done that to ourselves. Just think about that. His fear of aliens derives not from *actual* knowledge of aliens but from an actual knowledge of ourselves. Anytime a more *technologically* advanced [human] civilization has come upon a less advanced civilization, it did not bode well for the less advanced civilization. And that happened in North America with Europeans, South America with the Spanish,

Australia with the Brits—never boded well for the less technologically advanced civilization. His factual knowledge of that leads him to suspect that aliens would be exactly the same."

He goes on to say that he's not as skeptical as Hawking. DeGrasse Tyson doesn't think that all life forms in the Universe have the basal, primal, violent attitudes that we humans do as a species. And when Joe Rogan asks whether deGrasse Tyson believes that things advanced because of competition forcing things to be fairly ruthless, deGrasse Tyson's answer is revealing: "It has been argued that if you colonize, if you're a civilization that colonizes the Galaxy, that it's a self-limiting exercise. Why? Because, here you go. You ready? We start here on Earth. It's you and me, boy, all right? You take that planet, I take this planet. And now we both have offspring that are just like us. And we want more planets. All right, we reach a point where expansion is not possible because we are warring with ourselves to gain the territory that each other has obtained. So it has been argued sociologically that the very act of wanting to colonize is self-limiting against successful colonization of the Galaxy. Because to colonize the Galaxy it has to be done in an *organized* way. All right, you take this sector, I take this sector. But if I want territory and I want it now, and my kids want it now, I want *that* territory, not this other one, in fact I want it all. *That* kind of attitude breeds violence, it breeds war, intra-Galactic war. So it may be that the very kind of civilization that could *peacefully* colonize a Galaxy is not the kind of civilization that would colonize the Galaxy at all."

CLOSE ENCOUNTERS: WAS 'OUMUAMUA AN ALIEN CONTACT EVENT?

"It is pointless to worry about the possible malevolent intentions of an advanced civilization with whom we might make contact. It is more likely that the mere fact they have survived so long means they have learned to live with themselves and others. Perhaps our fears about extraterrestrial contact are merely a projection of our own backwardness, an expression of our guilty conscience about our past history: the ravages that have been visited on civilizations only slightly more backward than we."

—Carl Sagan, *Cosmos* (1980)

FANCY MEETING YOU HERE

In the 1976 movie *The Man Who Fell to Earth*, extraterrestrial contact arrives in the form of David Bowie playing a tragic, otherworldly hero, Thomas Jerome Newton, an alien who has come down to Earth in search of water for his drought-stricken home world. In the 2013 movie *Under the Skin*, contact comes in the form of Scarlett Johansson playing a seductress alien who lures men in Scotland into doomed sexual encounters. In the 2016 movie *10 Cloverfield Lane*, a third-person story (in contrast to its predecessor's found-footage format) presents an oblique account of an alien invasion in which evidence of the aliens isn't encountered until relatively late in the film—an alien bio-mechanical craft is seen floating in the distance.

Indeed, there seems to be as many *forms* of alien contact as there are movies about alien contact. Consider the plots to *Attack the Block*, *Midnight Special*, *Annihilation*, *District 9*, *Arrival*, or the 1977 Steven Spielberg classic, *Close Encounters of the Third Kind*. But would we recognize an alien contact event if we saw one? How would we know? Consider the following case from as recently as 2017.

THE VISITOR

In October of that year, Robert Weryk, a Canadian astronomer, was scanning images taken by the Pan-STARRS1 telescope when he suddenly spied something strange. Pan-STARRS1 sits atop Haleakalā, a huge volcanic peak on the island of Maui, and it sweeps the sky nightly, gathering data with Earth's highest-definition camera. The scope was designed and created to search for near-Earth objects, mostly asteroids whose trajectories swing our way into potential collision with our planet. Now, most asteroids travel at a mean speed of about ten miles a second, which is swifter than superman. And yet the speck of light that caught the eye of Robert Weryk was traveling almost five times that speed, at nearly two hundred thousand miles per hour (at that speed the speck of light could travel from New York to Los Angeles in under a minute).

Intrigued by this curious event, Weryk alerted fellow astronomers. They too began tracking the speedy speck from their respective observatories. And the more they spied the speck through their 'scopes, the more intriguing its behavior. The speck was small, was the consensus, about the size of a city block. As it careered through space, its luminosity changed so much, by a factor of ten, it simply *had* to have a very curious shape. The affair was beginning to sound like *Rendezvous with Rama*, a novel written by Arthur C. Clarke in 1973. In *Rama*, a rogue alien spaceship, first mistaken for an asteroid, is spotted in the vicinity of Jupiter, with the object's calculated speed at over one hundred thousand kilometers an hour.

What might be the shape of Weryk's object? In *Rendezvous with Rama*, the alien starship took the form of a cylinder, thirty-one miles by twelve miles. Robert Weryk's "interstellar object" was either long and skinny, like some kind of cosmic cigar, or round and flat, like a heavenly pita bread. Objects in our Solar System in orbit about the Sun are in elliptical orbits, such as the planetary bodies and comets whose orbits are so highly elliptical that they can spend hundreds or thousands of years out in the depths of the Solar System before they return to Sun at their perihelion. Weryk's speck of light, however, was zooming along in a straight(ish) line. This was new. This was something astronomers had never before seen. It truly did earn the description of "interstellar object." Like *Rama*, it seemed to be a visitor from far beyond the Solar System. Could it be that "ET" was passing through?

In the dry academic lexicon of the International Astronomical Union, Weryk's speck of light became known as 1I/2017 U1. For the rest of us, it was named 'Oumuamua, pronounced oh-mooah-mooah, a Hawaiian word meaning, roughly speaking, "scout." Was 'Oumuamua out scouting on behalf of an advanced and sophisticated alien race? After all, the entity seemed to race along as if propelled by an unknown force. Okay, sometimes comets get an extra impetus through cosmic space due to the gases they expel, which form those beautiful telltale tails. But 'Oumuamua didn't have a tail. And the telescopes trained on its journey found no sign of any of the by-products that usually come with outgassing, such as dust or water vapor.

WAS 'OUMUAMUA AN ET EVENT?

'Oumuamua is definitely a candidate for an unusual and possibly alien entity, just as portrayed by Arthur C. Clarke in *Rendezvous with Rama*. To merely add to the enigma, no new observations of 'Oumuamua are possible, as it's now way too dim and distant. And yet astronomers are still able to pore over the data already gathered.

Science is often a matter of eliminating hypotheses. And this is what the research scientists proceeded doing with 'Oumuamua. For one thing, the entity's weird motion couldn't be explained by a collision with another object. Nor could it be accounted for by interactions with the solar wind, that continuous flow of particles, especially electrons and protons, from the Sun that interacts with Earth's magnetic field and other interplanetary objects. One set of researchers considered the best explanation to be that 'Oumuamua was a miniature comet whose tail remained undetected due to its "unusual" chemical composition, which seems a rather exceptionalist argument. Another set of researchers suggested that the entity was made mostly of frozen hydrogen, another kind of mini-comet idea, which they felt explained the entity's curious shape. By the time it got to the Solar System, they argued, the mini comet had sublimated, like a popsicle in the Pinnacles Desert.

There were those who believed 'Oumuamua to be the creation of an alien civilization, just as in *Rendezvous with Rama*. In a math-heavy paper which was published in *The Astrophysical Journal Letters* twelve months or so after Robert Weryk's discovery, Harvard astrophysicists Avi Loeb and Shmuel Bialy suggested that 'Oumuamua's "non-gravitational acceleration" was best accounted for by concluding that the entity was not natural but constructed.

Perhaps 'Oumuamua was, like *Rama*, the equivalent of a cosmic ghost train, a deeply enigmatic alien artifact, seemingly floating abandoned in interstellar space as debris for some unsuspecting race to find. Maybe, should the discovering race be technological enough to land on 'Oumuamua, they might find it to be the Trojan Horse of an alien civilization. Or maybe it is a fully functioning alien probe, sent to our Solar System as reconnaissance, in which case we will see the invading armies of the Large Magellanic Cloud, let's say, storming the Earth's upper atmosphere before very long.

Maybe 'Oumuamua is space junk. Perhaps the entity could be explained as a large chunk of alien junk, zipping through the Galaxy, even if the chance of us discovering it would of course be ridiculously low. Believe it or not, a story similar to that of the theory that 'Oumuamua could be space junk has also been written about in sci-fi; fiction often seems to have run ahead of "fact." *Roadside Picnic*, written in 1972 by Soviet-Russian science fiction writers Arkady and Boris Strugatsky is an alien visitation tale with a difference.

The story is set in a post-visitation world, planet Earth, where there are now six mysterious Zones, regions of our globe that have been touched in some way by an alien visitation, some ten years past. The alien Visitors were never seen. But people local to the Zones reported loud explosions that blinded some and caused others to catch a kind of plague. Though the visit is thought to have been brief, around 12–24 hours, the half dozen Zones are full of mysterious phenomena, where strange events continue to occur. The location of the six Zones is not random. This discovery, made by fictional Nobel prize-winning physicist Dr. Valentine Pilman, is explained in a radio interview at the beginning of the book: "Imagine that you spin a huge globe and you start firing bullets into it. The bullet holes would lie on the surface in a smooth curve. The whole point (is that) all six Visitation Zones are situated on the surface of our planet as though someone had taken six shots at Earth from a pistol located somewhere along the Earth-Deneb line. Deneb is the alpha star in Cygnus." Now, no one is saying there are real-life Zones on today's Earth, or that the visitor is responsible for the COVID pandemic! But the idea of entities being shot in a straight line through space is an arresting one in the context of 'Oumuamua.

Naturally, Avi Loeb's extraterrestrial hypothesis for 'Oumuamua got a lot of flak. In defense of his theory, Loeb quoted Sherlock Holmes: "When you have excluded the impossible, whatever remains, however improbable, must be the truth." The hypothesis that Robert Weryk's entity was of an artificial origin was a story that girdled the globe at the speed of 'Oumuamua itself. To ward off the kind of personal and bitter attacks from narrow-minded astrophysicists that I have experienced myself, Loeb went to work on excluding the other theories of the nature of 'Oumuamua. Along with Thiem Hoang, a scientist at the Korea Astronomy and Space Science Institute, Loeb dropped the frozen-hydrogen theory into the dustbin of history. In another math-packed paper, they showed that it was the stuff of science fiction to suggest a mass of solid hydrogen floating around in deep space. And, even if a frozen chunk of hydrogen *did* manage to form, there's no chance it would survive in as large a form as 'Oumuamua after a long journey in interstellar space. As the pair put it, "assuming that hydrogen objects could somehow form, sublimation by collisional heating" would vaporize them before they had the chance to take off into space.

"AND YET IT DEVIATED"

Ditching the math-heavy journals, Avi Loeb nailed his academic colors to the mast in early 2021 with his book *Extraterrestrial: The First Sign of Intelligent Life Beyond Earth*. In an attempt to place himself on the right side of history, Loeb quotes the probably apocryphal words of Galileo at his trial in Rome in 1633. In response to accusations from the Inquisition that he was guilty of heresy for asserting that Earth circled the Sun, Galileo is meant to have said *"Eppur si muove"* (And yet it moves). Loeb seeks common cause with Galileo, who also had an astronomical establishment who wished to silence him. Loeb charges that, like the Inquisition who couldn't prove the Earth was not in motion, today's skeptics couldn't explain why 'Oumuamua had strayed from its expected path. "And yet it deviated," Loeb says.

What's the science behind Loeb's bold claim? It is, in a strong sense, the science of falsification. The Falsification Principle, proposed by twentieth-century Austrian-British philosopher Karl Popper, is a way of demarcating science from non-science. The Principle says that for a theory to be thought of as scientific, it must be able to be tested and conceivably proven false. For instance, the hypothesis that "all swans are white" can be falsified by seeing a black swan.

In his book *Extraterrestrial*, Loeb goes through the counter theories for 'Oumuamua and shows each of them to be false. In short, the sole way to understand 'Oumuamua's curious acceleration, without sticking to the theory of outgassing, which can be falsified because outgassing is undetectable, is to consider that the entity was propelled by solar radiation, that is, photons ricocheting off its surface. And the sole way the entity could possibly be propelled by solar radiation is if it were exceptionally thin, around a millimeter, and with a low density and a relatively large surface area. An entity such as that would work as a sail, albeit one powered by light, not wind. But the cosmos doesn't produce sails; creatures like us do. Therefore, Loeb concludes "'Oumuamua must have been designed, built, and launched by an extraterrestrial intelligence."

Legendary American astronomer and science communicator Carl Sagan is often quoted as saying "extraordinary claims require extraordinary evidence." Like Loeb, Sagan too was open-minded about the prospect of alien contact in the future as well as in the distant past. By this Sagan standard of extraordinary evidence, Loeb's case probably still has a lot to answer for. And yet Loeb's reply is that "extraordinary conservatism keeps us extraordinarily ignorant." In other words, if we deny the possibility that 1I/2017 U1 is an alien entity, "whole new vistas of exploration for evidence and discovery open before us," as Loeb puts it. In sticking out his neck and publishing his thoughts, Loeb has certainly risked his reputation. And yet, there may be a whole new generation of stargazers who now look to the skies and wonder if we might hear more from 'Oumuamua, or other ways in which alien contact might come in the near future.

CLOVERFIELD AND *CONTACT*: WHAT IS THE MOST LIKELY FORM OF ALIEN CONTACT?

Executive: "We must confess that your proposal seems less like science and more like science fiction."

Ellie Arroway: "Science fiction. Well, you're right, it's crazy. In fact, it's even worse than that; nuts. You wanna hear something really nutty? I heard of a couple guys who wanna build something called an 'airplane,' you know you get people to go in, and fly around like birds, it's ridiculous, right? And what about breaking the sound barrier, or rockets to the Moon, or atomic energy, or a mission to Mars? Science fiction, right? Look, all I'm asking, is for you to just have the tiniest bit of vision. You know, to just sit back for one minute and look at the big picture. To take a chance on something that just might end up being the most profoundly impactful moment for humanity, for the history . . . of history."
—James V. Hart and Michael Goldenberg, *Contact* (1997)

THE NAZI ALIENS OF SPACE-TIME

Imagine aliens beaming an image of a swastika to your science lab. That would be very much worse than a bad hair day. It sounds like the kind of thing the Daleks would do. As extraterrestrials go, these *Doctor Who* aliens are closest to the Nazis in attitude. They believe every other race inferior and are hell-bent on conquering the Universe. Dalek-creator Terry Nation actually based his invention *on* the Nazis. And when you watch the Daleks armed with that knowledge, their behavior makes a lot more sense. In the very first two Dalek stories, the Daleks even make the Nazi salute. On their home planet, the Daleks gradually usurped the power of the legitimate government with the support of a quasi-military organization that wear black uniforms bearing insignia reminiscent of lightning bolts and who salute each other by raising a hand, palm outward, and clicking their heels together. I rest my case.

In Robert Zemeckis's 1997 movie adaptation of Carl Sagan's *Contact*, a terrestrial science team discovers a video hidden in an extraterrestrial signal. The video just happens to be Adolf Hitler's opening address at the 1936 Summer Olympics in Berlin, Germany, whose opening frames focus in on the swastika symbol. The team soon realize that the 1936 Olympics coverage would have been the first signal strong enough to leave Earth's ionosphere, reach Vega (the source of the alien signal), and be transmitted back. In contrast, the 2008 Matt Reeves movie *Cloverfield* presents an alien invasion as seen from below, not above. There are few elites in *Cloverfield*. No establishment scientists, military, or political personnel. Instead, we get a found-footage format, recorded from ordinary people's personal camcorders, which presents an oblique account of the alien invasion. If or when contact comes, which of the two scenarios is the most likely: *Cloverfield* or *Contact*?

THE GREATEST ALIEN CONTACT MOVIES

To help think about this question, it's worth considering how much our minds are made up by the movies. Movie history has spawned far more alien encounters than you realize. The famous among them have ranged from the iconic imagery of the Devils Tower and the sound of the five-note musical phrase used to communicate with the aliens in Steven Spielberg's

1977 seminal film *Close Encounters of the Third Kind* to the stomach-hitchhiking alien species in Ridley Scott's 1979 movie classic *Alien*.

As this is the movies, the aliens usually make their appearance in the most dramatic of fashions. In Denis Villeneuve's 2016 science fiction film *Arrival*, we first see twelve extraterrestrial spacecraft hover over various locations around the Earth. In Roland Emmerich's 1996 film *Independence Day*, it's a similar story. We see the event through the eyes of Will Smith's character as he sleepily picks up the daily newspaper from his drive, slowly notices all the neighbors looking up to the skies, spots a chopper scurrying overhead, then finally claps eyes on the real story: an enormous alien mothership has arrived in Earth's orbit and has sent out assault fortress saucers, each fifteen miles wide, that have taken up threatening positions over Earth's major cities.

THE BIRTH OF STRATEGIC ALIEN INVASION

This popular portrayal of alien invasion, by strategically placing alien ships over certain positions on Earth's surface, has its most famous original expression in Arthur C. Clarke's 1953 novel *Childhood's End*. Clarke had already written a number of short stories on aliens. The influential *The City and the Stars* portrays humanity confronted with extraterrestrial cultures and intelligences "he could understand but not match, and here and there he encountered minds which would soon have passed altogether beyond his comprehension." In *Childhood's End*, Clarke developed the myth of contact through alien invasion. The story's "Overlords" are benevolently responsible for guiding humanity to an even greater intelligence, the "Overmind." The Overlords exact an end to poverty, ignorance, war, and government. But there is a price to pay. It is a preparation for the final destiny of humanity, and Earth children are to be sacrificed and united within the collective of the Overmind. Clarke's novel is referenced in one of David Bowie's most famous songs, "Oh! You Pretty Things," a track on his 1971 album *Hunky Dory*. It includes the lyrics: "I look out my window what do I see/A crack in the sky and a hand reaching down to me/All the nightmares came today/And it looks as though they're here to stay . . . Look out at your children/See their faces in golden rays/Don't kid yourself they belong to you/They're the start of a coming race/The Earth is a bitch/

We've finished our news/Homo Sapiens have outgrown their use/All the strangers came today/And it looks as though they're here to stay."

Clarke got this idea of the strategic placement of alien ships during World War II. He and a colleague had driven to the outskirts of London and looked back toward the city center. What they saw stunned them: a menacing host of barrage balloons floating over the city. Clarke later recalled that his earliest idea for *Childhood's End* originated with this scene, with the giant balloons becoming alien ships in the novel. Yet, despite Clarke's development of the persuasive and dramatic myth of alien invasion, the remote kind of extraterrestrial encounter portrayed in *Contact* is far more likely.

JOHN WAS TRYING TO CONTACT ALIENS: ARE WE ALONE?

"Sometimes, taking the course that I have in my life, the path is like maybe a lonely mountain road to some higher-elevation peaks to see the view, to check out something most people don't see. So you tend to go it alone more. To find someone that's on the wavelength that I am on, and be able to share my life with that person in any degree or way, is nearly impossible, although I believe it exists. I believe for everyone, there is someone."
—John Shepherd, *John Was Trying to Contact Aliens* (2020)

JOHN WAS TRYING TO CONTACT ALIENS

The whole history of SETI revolves around a number of fascinating puzzler questions. If you were lucky enough to be able to talk to aliens, what on Earth would you say? How would you make yourself known and understood by a culture that was so truly alien? And, given the cosmic chasm between our respective cultures, terrestrial and extraterrestrial, what kind of context is universal?

The 2020 Netflix documentary *John Was Trying to Contact Aliens* tells the tale of John Shepherd, an American amateur astronomer, who has been trying to contact aliens for over thirty years. His chosen method of communication? Beaming music into space with his state-of-the-art broadcasting kit. For John Shepherd, those fascinating puzzler questions are far more than hypothetical. They are opportunities with potentially rewarding solutions. Through the decades of the seventies, eighties, and nineties, John Shepherd carried out ambitious, one-man attempts to talk to aliens from his Michigan home. The universal cosmic context, Shepherd resolved, was music. Whether reggae, Afro-beat, or jazz, he beamed music out into deep space through a complex collection of tech which made it look like his cottage was slowly morphing into a starship.

The Netflix documentary, a mere sixteen minutes long, won the short film jury award at the Sundance Film Festival in January 2020. In the movie, we find Shepherd at home, still in Michigan, and still infatuated with the potential of outer space and alien contact. Surrounded by the type of tech that seems more in keeping with a spoof sci-fi movie, Shepherd clearly finds some fulfillment, up to a point, in his tiny part of our planet. After decades of developing the art of his science, Shepherd triggers the tech, sitting before a wall of knobs and dials. Like a Greek philosopher or a guru, Shepherd is bearded. Like an aging hippy, his hair hangs loose in a ponytail. And like a hermit of his Michigan hollow, he is one part timid and the other part zealous as he turns yet another dial, which generates a wave pattern, a thrum of electric hum, and the search for ET.

EXPERIMENTAL ROCK

Back in 1977, NASA launched the Voyager spacecrafts to explore the cosmos. They sent along the Golden Record, which carried with it music from around the world, as a snapshot portrait of human culture, should smart aliens ever find it. But, Shepherd wondered, what if aliens preferred the radio? And so, from 1971 to 1998, he probed that question with almost religious devotion. Adopted and raised by his grandparents, Shepherd became a local media star in the era when sci-fi movies met CGI for the first time. Shepherd's mission to contact extraterrestrial beings began at an early age, and his faith abides: "My interest is in finding out the unknown, and the unknown is just that—unknown. And you search, and you continue searching, because of your desire, because you know there's something there."

Shepherd's alien contact hypothesis is this: Music is universal. It's a way of inviting aliens to connect, to engage their intellectual and cultural curiosity, and in so doing draw them close enough to Earth so that they could be studied. True, it's a long shot. But the project is a child of the times Shepherd was living through. When Shepherd's project began in 1971, experimental rock was in full flow. By the late sixties, music artists had begun creating more complex compositions through new tech such as multitrack recording. Artists aimed to liberate and innovate. And some of the subgenre's distinguishing features were groundbreaking instrumentation, unorthodox structures and rhythms, and an unwritten disregard for commercial aspirations.

Experimental rock was the very subgenre that Shepherd chose for his alien contact work, music that was boundary pushing and globally minded. Albums like *Mysteries* by jazz pianist Keith Jarrett, *Give Me Power!* by roots reggae group the Itals, *Future Days* by the krautrock band Can, and a compilation of late seventies music from Afro-beat pioneer Fela Kuti. All these musicians make cameos in the Netflix documentary when Shepherd puts his extensive record collection into overdrive. He was specifically drawn to instrumental music for his alien broadcasts, which he planned into hourly programming blocks. His radio shows were diverse in tone and theme and often delved deep into subgenres such as early electronic music like Kraftwerk and Tangerine Dream, exploratory jazz, and percussive music of Indonesian gamelan ensembles.

PROJECT STRAT AND PROJECT HUMAN

Armed with his own particular version of the extraterrestrial hypothesis, Shepherd began, along with the help of his grandfather, to build banks of electrical equipment—boxy towers bejeweled with screens and dials. His one-man mission he named "Project Strat" (Special Telemetry Research and Tracking), and the tech soon took up whole rooms of the house. An old photograph captures the surreal scope of Shepherd's project. In the foreground, a young hippy-looking Shepherd loads a music cassette into an apparatus which looks like the kind of kit you get in a physics lab, but set out on what is clearly a normal dining table. Meanwhile, in the background, a children's doll sits on top of a piano, as if surveying Shepherd's efforts, and the trappings of a typical 1970s home suggest there's nothing unusual to see here. Beaming music out to aliens in deep space is an everyday activity.

The Netflix documentary is a fine example of how the search for alien life is a cultural, as well as scientific, inquiry. The documentary's director, Matthew Killip, explained that "if you make a film about someone trying to contact aliens, there's an in-built narrative problem, which is that they don't contact aliens." Indeed, the same in-built narrative problem is true of this very book, and the entire field of extraterrestrial studies! Given that the basic elements of the scientific method are to generate a hypothesis, test its premise, find (or not) empirical evidence, and then come up with conclusions, the academic field of extraterrestrial studies is sorely lacking.

The documentary film shows that Shepherd's lifelong interest in contacting aliens in deep space is deeply romantic, and profoundly emblematic of the human condition over the last few hundred years, especially the last century or so. Shepherd's story is far more universal than a guy keeping thousands of dollars of radio and electrical equipment at his grandparents' house might seem on first inspection. As Killip says, "We're all sort of sending out a message hoping that someone else will pick it up and understand us and understand who we are. We're all trying to make contact." Shepherd's tale brings "the search for love, or a place in the world, or a partner that recognizes you, or a family that recognizes you, into a kind of cosmic context." As Shepherd says partway through the documentary, "Sometimes, taking the course that I have in my life, the path is like maybe a lonely mountain road to some higher-elevation peaks to see the view, to check out something most people don't see. So you tend to go it alone more."

John Was Trying to Contact Aliens is far from being a long documentary. And yet it packs a punch. The film vividly portrays Shepherd's search for terrestrial love. How, when younger, Shepherd discovered that he was gay. How his hermit existence in a relatively remote town, and his passion with space, meant the chance of meeting a life partner seem as remote as alien contact. In an older interview he admits, "To find someone that's on the wavelength that I am on, and be able to share my life with that person in any degree or way, is nearly impossible, although I believe it exists. I believe for everyone, there is someone."

Though no alien contact has happened (yet), Shepherd found a life partner. A man, also called John, long-haired and bearded, who now lives with Shepherd in Michigan. Though the space radio broadcasts have stopped in recent years, his sense of fascination remains. Thirty years of trying alien contact by engaging ETs with "cultural music" left him with "little hard data." But, as Shepherd says in the documentary, his time spent on contact was creative and connective enough: "It filled my life. It gave it something—meaning."

ALIENS AND ANGELS

The idea of life beyond this Earth is an ancient one. Surveys of public opinion today suggest that most people think of SETI as a space age activity, one born of the first attempts of Drake and co, back in 1961. But, as this book has shown, the belief in, and the search for, alien life stretches way back into the Classical past. These days we tend to separate a belief in aliens from a belief in angels. But is that true for everyone? For most of the two-and-a-half-thousand-year history we've looked at in this book, the "heavens" were the realm of the angels and the gods, supernatural beings who inhabited the supra-lunary aether, that realm beyond our sub-lunary domain. As we have seen, the sub-lunary sphere, from the Earth to the Moon, was the only part of the ancient cosmos considered to be subject to the horrors of change. Beyond the Moon, the supra-lunary sphere, all was unchanging, the domain of the gods and their agents.

As *John Was Trying to Contact Aliens* shows, although the search for alien life is solidly seated in the realm of science, its intimate relationship with science fiction and the religious dynamic of the search still abides, still lurks beneath the surface. Like Shepherd, many folks are comforted by the idea that there are advanced and sophisticated beings in the heavens, watching over us, and who may someday intercede in our petty affairs to save us from ourselves. And, for many, an obsession with alien life and contact is often an expression in pseudo-tech language of ancient beliefs in the supernatural, which has been part and parcel of many cultures through the ages.

One can even find evidence of contact-like reports in antiquity. Take, for example, the many tales in the Christian bible which tell of angels coming down from the sky and ascending (abducting) with humans into heaven, or of flying chariots. Perhaps the most famous of these stories is that of Ezekiel, a central protagonist of the Book of Ezekiel in the Hebrew bible. In Judaism, Christianity, and Islam, Ezekiel is acknowledged as a Hebrew prophet. The Book of Ezekiel describes an "alien" encounter with four flying wheel-shaped crafts "full of eyes," that "turned as they went," and from which stepped "the likeness of a man." The account reads like the script of a twentieth-century alien contact movie, and yet it's normally read in a purely religious way. Indeed, Carl Jung, Swiss psychoanalyst and founder of analytical psychology, has suggested that the flying saucers of modern science fiction are simply the most recent manifestation of the kind of archetypal symbols which have appeared in religious dreams and visions since antiquity.

"I HAVE SPOKE WITH THE TONGUE OF ANGELS"

One can easily see why some people argue that, at least in part, the search for aliens feels like a kind of religious quest. And the interest in the search among the general public, as with John Shepherd, springs from the need to seek a wider, more cosmic context for our contemporary lives, beyond what our Earthly existence seems to provide. As God has moved to the margins and shadows of everyday life for many people, the angels have been replaced by aliens.

Though many scientists, like myself, are self-proclaimed atheists, nonetheless some sense of a spiritual quest may be part of their crusade to make contact. Consider, once more, the celebrated SETI campaigner Carl Sagan. As we have seen, Sagan's 1985 novel *Contact* is an account of a successful alien encounter. After the scientists receive the, at first alarming, swastika signal, they build a spacecraft from the alien design they received in the signal and venture deep into the Galaxy. At contact, the alien beings kindly use familiar landscapes and familiar forms to ensure that initial contact is easier for our human envoy, played by Jodie Foster. At the same time, we learn some profound things about the very nature of the Universe; that this journey is just humanity's first step to joining other spacefaring species in the cosmos, and that we have become members of a Galactic Club of sentient species.

And yet the *subtext* of Sagan's story is the idea of a *designed* cosmos. The aliens suggest how the detail of this design is coded into the very structure of the cosmos. It's easy to interpret this helping heavenly hand as angelic; with the aliens as agents of the void, messengers of a pantheistic Universe, designed for sentient beings to thrive once they come of age.

EARTH AS HALFWAY HOUSE

Way back with Aristotle's ancient two-tier geocentric cosmology, the Earth had been something of a halfway house. As we saw earlier in this book (page 6), Aristotle placed the Earth, mutable and corruptible, at the center of a nested system of crystalline celestial spheres, from the sub-lunary to the sphere of the fixed stars. A similar vision of aliens as a kind of halfway house to the heavens is a regular theme in science fiction. And this theme is sometimes picked up by prominent scientists in nonfiction too. Consider the British physicist Fred Hoyle. In his book *The Intelligent Universe*, Hoyle explored the idea that life didn't originate on Earth, but extends throughout the cosmos. According to Hoyle, terrestrial life began when a shower of microorganisms, biological precursors for life, hit our atmosphere and seeded our virgin planet. Ring any bells? Yes, it's a similar plot to Ridley Scott's *Prometheus* and the *Star Trek: The Next Generation* episode "The Chase," among many others.

What's the mechanism for the generation of these organisms? Hoyle dumps the idea of a random origin for these organisms and instead hints at the existence of advanced sentient beings, out there in deep space, who deliberately contrived to create in our cosmic backyard the conditions necessary for carbon-based bipeds. Not content with what sounds like a modern-day version of angels, Hoyle then goes on to describe a far more potent and powerful super-intelligence, an entity which directs the laws of intelligent design from some timeless vantage point of the infinite future. Such theories, though they sound so fantastic they're even beyond the scope of most science fiction films, strike a deep chord in the human psyche. And, from this perspective, the search for aliens feels like a continuation of a long-standing spiritual quest, as well as a scientific mission.

This perspective also helps us better understand the role of alien fiction in science. Ever since the early days of the Scientific Revolution, alien fiction was launched alongside the new physics. They marked that paradigm shift of the old and cozy cosmos of Aristotle into the new Universe of today; decentralized, infinite, and alien. The old cosmos had the stamp of God and angels about it, and the heavens spoke of evidence of God's glory. But alien fiction was a response to the new and Godless Universe. It dealt with the shock created by the discovery of our questionable position in an empty cosmos. Earth was no longer at its center, nor was it made of a unique material only to be found on terra firma. Alien fiction tried to fill the void, to make human what is alien. After all, the truth could be just as Loren Eiseley said in *The Immense Journey*: "In the nature of life and in the principles of evolution we have had our answer. Of men elsewhere, and beyond, there will be none forever."

INDEX